International Conference on Materials Research and Innovation (ICMARI 2021)

IOP Conference Series: Materials Science and Engineering
Volume 1234

Bangkok, Thailand
15 – 17 December 2021

Editors:

Wirasak Smitthipong
Saree Phongphanphanee
Jirasak Wong-ekkabut

ISBN: 978-1-7138-4691-8
ISSN: 1757-8981

Printed from e-media with permission by:

Curran Associates, Inc.
57 Morehouse Lane
Red Hook, NY 12571

CURRAN ASSOCIATES INC. proceedings.com

Some format issues inherent in the e-media version may also appear in this print version.

Institute of Physics
Dirac House, Temple Back
Bristol BS1 6BE UK

Phone: 44 1 17 929 7481
Fax: 44 1 17 920 0979

techtracking@iop.org

Additional copies of this publication are available from:

Curran Associates, Inc.
57 Morehouse Lane
Red Hook, NY 12571 USA
Phone: 845-758-0400
Fax: 845-758-2633
Email: curran@proceedings.com
Web: www.proceedings.com

TABLE OF CONTENTS

IOP Conf. Series: Materials Science and Engineering 1234 (2022) 011001 doi:10.1088/1757-899X/1234/1/011001

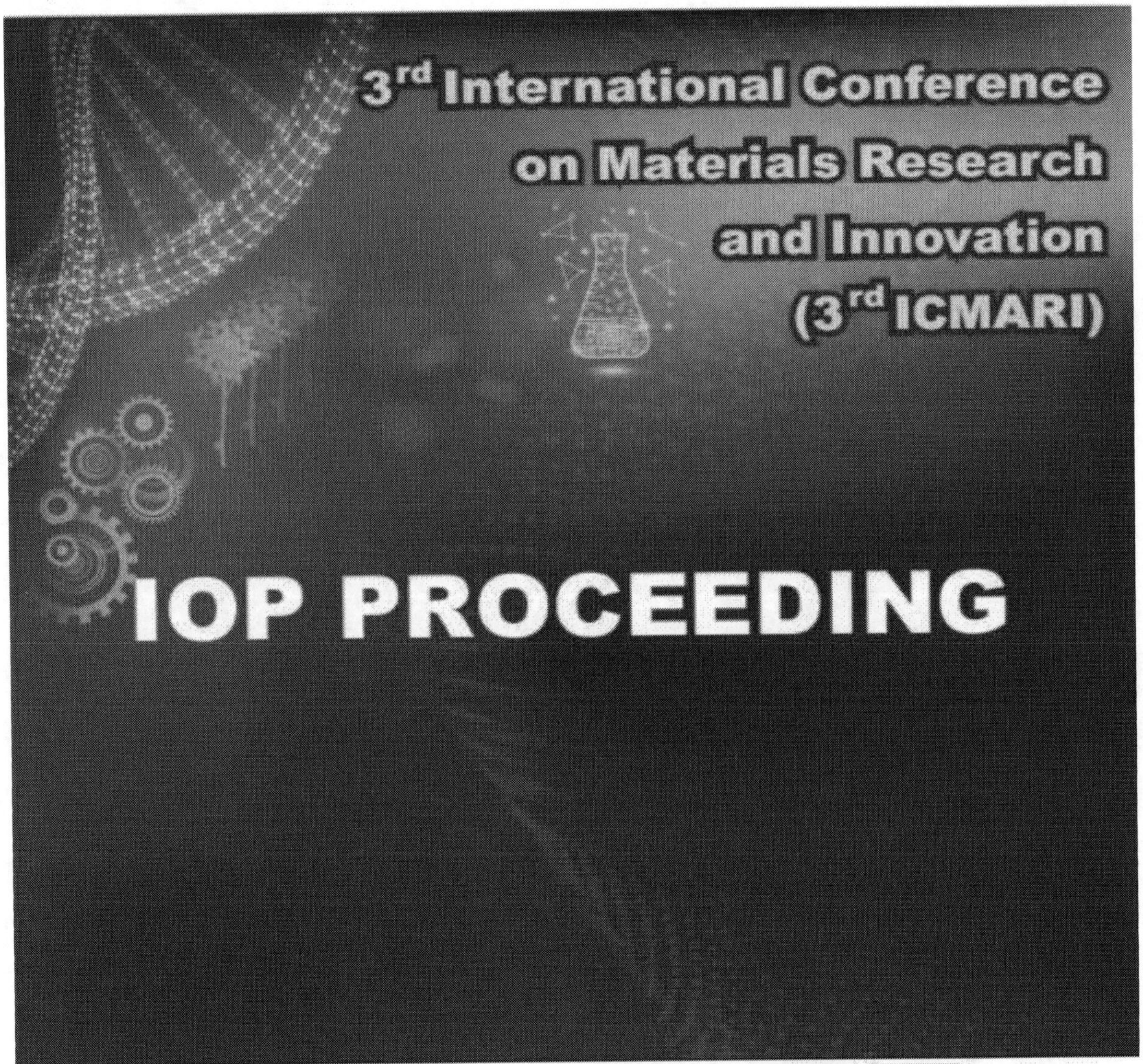

Preface

The 3rd International Conference on Materials Research and Innovation (ICMARI 2021) was held on 15th – 17th December 2021, Bangkok, Thailand. The conference was conducted in frontier research on materials research and Innovation including Rubber and Polymeric Materials, Bio- and Green-materials, and Special Advanced Materials. This 3rd ICMARI was announced to focused on the Bio-Circular-Green Economic Model or BCG. The BCG has been applied by the research community and promoted by the Thai government as a new economic model for inclusive and sustainable growth, this BCG model also conforms with the UN Sustainable Development Goals (SDGs). The 3rd ICMARI was a great opportunity for all participants to exchange knowledge and strengthen the research collaboration not only in ASEAN but also around the world. We hoped that the participants are fruitful by plenary and invited speakers from novel knowledge to new applications in the field of materials research and innovation.

Thank you all for participating,

ICMARI 2021 Organizing Committee

Organizing Committee

Asst. Prof. Wirasak Smitthipong (Chairman)

Committee (Inside KU)

Asst. Prof. Saree	Phongphanphanee
Asst. Prof. Nattasamon	Petchsang
Asst. Prof. Chanapa	Kongmark
Asst. Prof. Decha	Dechtrirat
Asst. Prof. Pongthep	Prajongtat
Assoc. Prof. Sutee	Boonchui
Assoc. Prof. Jirasak	Wong-ekkabut
Asst. Prof. Chomdao	Sinthuvanich
Asst. Prof. Wirunya	Keawwattana
Assoc. Prof. Kiadtisak	Saenboonruang
Assoc. Prof. Patthra	Pengthamkeerati
Asst. Prof. Jukkrit	Mahujchariyawong
Assoc. Prof. Nanthiya	Hansupalak
Asst. Prof. Naraphorn	Paoprasert
Assoc. Prof. Prakit	Sukyai
Asst. Prof. Parames	Kamhangritirong
Ms. Siriwan	Nerdnoi

Committee (outside KU)

Mr. Payab	Namprasert
Mrs. Pannee	Jirabunpaisit

Secretary team

Ms. Ratchadathip	Purnajoti
Dr. Supitta	Suethao
Mr. Kwanchai	Buaksuntear
Ms. Siriwan	Jansinak
Ms. Phakamat	Lim-arun
Ms. Noppawan	Tundiew
Ms. Kanyapat	Wongpul
Ms. Kesinee	Panmanee
Mr. Peerayut	Kunklang
Ms. Piyanut	Promkaowthong
Ms. Haruthai	Klommueang
Ms. Nantinee	Choosang
Mr. Watcharapong	Wilaiwong

IOP Conf. Series: Materials Science and Engineering 1234 (2022) 011001 doi:10.1088/1757-899X/1234/1/011001

Editors of Proceeding

Asst. Prof. Wirasak	Smitthipong	Chair Editor
Asst. Prof. Saree	Phongphanphanee	Editor
Assoc. Prof. Jirasak	Wong-ekkabut	Editor

International Technical Committee

Prof. Herbert	Waite	University of California, Santa Barbara	USA
Prof. Sadhan C.	Jana	University of Akron	USA
Assoc. Prof. Karine	MOUGIN	Institut de Science des Materiaux de Mulhouse	France
Assoc. Prof. Kheng-Lim	Goh	Newcastle University in Singapore	Singapore
Asst. Prof. Parames	Kamhangritirong	Kasetsart University	Thailand
Asst. Prof. Wirasak	Smitthipong	Kasetsart University	Thailand
Prof. Arkadiusz	Chworos	Polish Academy of Sciences	Poland
Asst. Prof. Nattasamon	Petchsang	Kasetsart University	Thailand
Asst. Prof. Chanapa	Kongmark	Kasetsart University	Thailand
Asst. Prof. Saree	Phongphanphanee	Kasetsart University	Thailand
Dr. Darshil U.	Shah	University of Cambridge	UK
Assoc. Prof. Norio	Yoshida	Kyushu University	Japan
Prof. Katsura	Nishiyama	Meijo University	Japan

Peer Review Statement

All papers published in this volume have been reviewed through processes administered by the Editors. Reviews were conducted by expert referees to the professional and scientific standards expected of a proceedings journal published by IOP Publishing.

1. Type of peer review: Single Anonymous
2. Conference submission management system: Morressier
3. Number of submissions received: 39
4. Number of submissions sent for review: 39
5. Number of submissions accepted: 37
6. Acceptance Rate (Submissions Accepted / Submissions Received × 100): 94.87 %
7. Average number of reviews per paper: 2.00
8. Total number of reviewers involved: 20
9. Contact person for queries:
   Name: Asst. Prof. Wirasak Smitthipong; Asst. Prof. Saree Phongphanphanee; Assoc. Prof. Jirasak Wong-ekkabut
   Affiliation: Kasetsart University
   Email: fsciwssm@ku.ac.th

* means value has been edited

Theophylline extended-release monolithic matrix comprising natural rubber latex as binder

P Chaiya[1] and T Phaechamud[2*]

[1] School of Pharmacy, Walailak University, Tha sala, Nakhon Si Thammarat 80160, Thailand
[2] Department of Pharmaceutical Technology, Faculty of Pharmacy, Silpakorn University, Nakhon Pathom 73000, Thailand

*Corresponding author: phaechamud_t@su.ac.th

Abstract. A major component of polymer in natural rubber latex (NRL) obtained from Hevea brasiliensis consists of poly-*cis*-1,4-isoprene. Poly-*cis*-1,4-isoprene exhibited interesting physical properties suitable for possible use as a binder in pharmaceutical solid dosage forms such as monolithic matrix tablet. The aim of this study was to study the feasibility of using NRL as the binder in theophylline anhydrous (THE)-incorporated monolithic matrix tablet in comparison with polyvinyl pyrrolidone K-30 (PVP-30). Physical properties of granules and tablets fabricated with wet granulation and tableting process, *in vitro* drug release test and release kinetic model were investigated and compared with those prepared using PVP-K30 as a binder. THE-incorporated monolithic matrix tablet was successfully fabricated with wet granulation and tableting process using NRL or PVP-K30 as binder. Granules and tablets of monolithic matrix tablet using NRL as a binder exhibited similar physical properties to those of tablet prepared using PVP-K30 as a binder. In addition, the controlled THE release with diffusion mechanism were presented in both formulations. Surprisingly, the addition of sodium bicarbonate in THE monolithic matrix tablets employing NRL as a binder could also prolong the drug release.

1. Introduction

Natural rubber latex (NRL), a milky fluid obtained from Hevea brasiliensis, is a colloid of poly-*cis*-1,4-isoprene particles dispersed in an aqueous serum [1]. This polymer possesses interesting physical properties such as high elasticity and tensile strength. Therefore, NRL may have potential to be applied as a binder in solid dosage forms such as monolithic matrix tablet. This study focused on using NRL as binder for preparing monolithic matrix tablets and compared with polyvinyl pyrrolidone K-30 (PVP K-30), the commonly used binder in pharmaceutical industry. In this study, theophylline anhydrous (THE), a non-selective phosphodiesterase-4 (PDE-4) inhibitor mainly used in treatment of asthma and chronic obstructive pulmonary disease (COPD), was used as the model drug [2]. The aim of this study was to study the feasibility of using NRL as binder in monolithic matrix tablet fabricated with wet granulation and tableting process, and to investigate the physical properties of granules and tablets. *In vitro* drug release test and release kinetic model were also investigated.

2. Experimental

2.1 Materials

The concentrated natural rubber latex (NRL) containing 60% dry rubber content (stabilized with 0.69% ammonia) was procured from Thai Rubber Latex Corporation (Thailand) Public Company Limited, Thailand. Theophylline anhydrous (THE) (Batch No. A000725), sodium bicarbonate (BICARB) (Batch No. AF310196, Ajax Finchem, Seven Hills, Australia), magnesium stearate (MgSt), dibasic calcium phosphate (DCP) and polyvinyl pyrrolidone K-30 (PVP-K30) were supplied by P.C. Drug center CO., LTD., Bangkok, Thailand.

2.2 Granules and tablets preparations

In this study, three different THE formulations were prepared by the wet granulation method, and their granules were compressed into a monolithic matrix tablet. NRL (Non-Bicarb NR matrix tablet) or PVP-K30 (PVP tablet) were used as binders in THE formulations. In addition, sodium bicarbonate was used as an effervescent base in Bicarb NR matrix tablet. THE and DCP were gradually mixed with porcelain mortar and pestle. Then binder solutions (NRL or PVP-K30 in purified water) were incorporated until wet mass was obtained. Wet mass was sieved through stainless steel sieve no. 16 and dried in hot air oven at 50°C for 4 h. Dried granules were sieved through stainless steel sieve no.20 and MgSt was added at concentration of 1%w/w of the obtained dried granule and mixed for 3 min with a plastic bottle. Dried granules were compressed into 200-mg monolithic matrix tablet by hydraulic press (Caver press, WI) at compression force of 4 tons with dwell time of 10 sec. Each monolithic matrix tablet contained 100 mg THE.

2.3 Evaluations

Physical properties and surface topography of granules and tablets
Tapped and bulk density of dried granules were tested with tapped density tester (PHARMA TEST, USA) then, calculated and expressed as compressibility. In addition, the moisture content and angle of repose were also tested. In case of tablets, 10 tablets of each formulation were tested for their diameter and hardness with a hardness tester (PHARMA TEST, USA) and thickness with a thickness tester (Teclock, Japan). The wettability was measured using a goniometer (First Ten Angstroms, FTA 1000, USA). DI water and formamide were chosen as test solvent. The obtained contact angle (n=3) was recorded and then, surface free energy (SFE) of each formulation was calculated by Wu harmonic method. In addition, the dried granules of all formulations were photographed with scanning electron microscope (SEM) (Cam Scan MX-2000, UK) at appropriate magnification.

Water sorption and erosion studies
Water sorption and erosion of all formulations were investigated as follows. The weight of THE tablet (W_0) was determined before the test, then the tablet was immersed into the glass bottle containing 100 mL of HCl buffer pH 1.2 and rotated at 50 rpm at 37°C using shaking incubator (Model SI4, Shel Lab, Cornelius, USA) for 24 h. Then monolithic matrix tablets were withdrawn, blotted to remove excess water with filter paper, and the wet weight (W_1) were recorded. After that, they were placed in a hot air oven at 80°C until the tablets weight did not change, and the dry weights (W_2) were obtained. The water sorption and erosion characteristics of the THE monolithic matrix tablets were expressed in terms of water uptake (%) and mass loss (%) (n=6).

$$\text{Water uptake (\%)} = \frac{W_1 - W_2}{W_2} \times 100 \quad (1)$$

$$\text{Mass loss (\%)} = \frac{W_0 - W_2}{W_0} \times 100 \quad (2)$$

***In vitro* drug release study and kinetic of drug release**
The *in vitro* drug release was investigated in HCl buffer pH 1.2 with dissolution apparatus (Prolabo, France) using paddle method at 37°C and rotational speed of 50 rpm. and THE content was analysed using a UV-vis spectrophotometer (Perkin-Elmer, Germany) at 269 nm (n=6). After release test, Bicarb

NR tablet was dried with lyophilizer (Labconco, MO, USA.); then photographed with SEM (Cam Scan MX-2000, UK). The *in vitro* drug release profiles were fitted with four different kinetic models including zero order, first order, Higuchi's model and power law to explain the mechanism of THE release.

3. Results and discussion

Wet granulation of THE monolithic matrix tablet was successfully fabricated using PVP-K30 or NRL as the binders in formulation. Dried granules of all formulations exhibited acceptable compressibility. Angle of repose and moisture content were higher in Bicarb NR matrix tablet (data not shown). This result was owing to hygroscopic properties of sodium carbonate decomposed from reaction between moisture in air and sodium bicarbonate in formulation, affecting the flowability of dried granules. All THE matrix tablets had good physical appearance and high uniformity in thickness, diameter, hardness and weight. The hardness of Non-Bicarb NR monolithic matrix tablet using NRL as a binder was slightly higher than that of PVP-K30 tablet (data not shown). Granules and tablets of monolithic matrix tablet using NRL as a binder exhibited similar physical properties to those of tablet that used PVP-K30, a commonly used commercial binder in pharmaceutical industries.

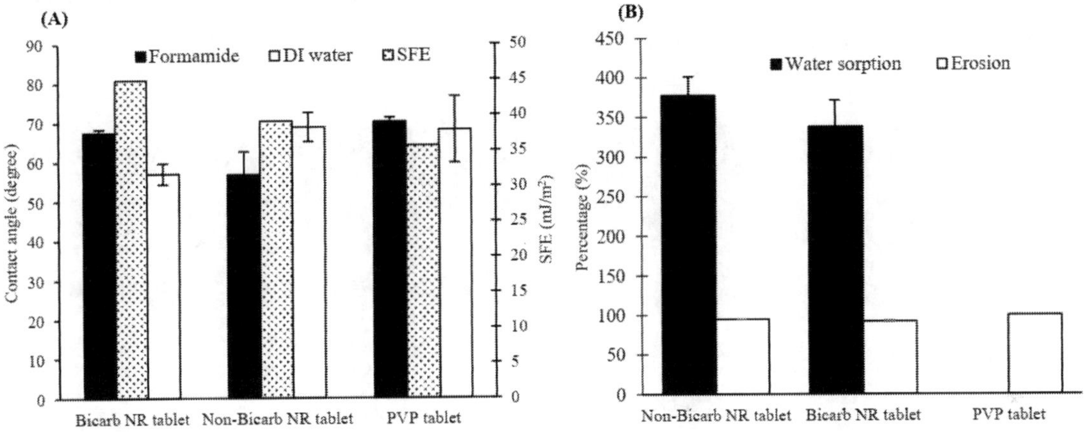

Figure 1. wettability (n=3) (A) and water sorption and erosion properties (n=6) (B) of THE tablets.

The wettability of THE matrix tablet determined from contact angle and SFE are shown in Figure 1A. The contact angle of DI water on Bicarb NR matrix tablet was lower than other two matrix formulations, indicating that it had more hydrophilicity. For SFE, Bicarb NR matrix tablet showed the highest SFE which was consistent with contact angle results. In case of water sorption and erosion test, PVP tablet completely dissolved after immersed in HCl buffer for 24 h. Thus, its water sorption could not be calculated whereas the erosion property of this formulation was assumed as 100%. Both Bicarb NR and Non-Bicarb NR tablets showed the same erosion property at about 95% of initial weight (Figure 1B). Water sorption of Bicarb NR tablet was slightly, but not significantly, lower than that of Non-Bicarb NR matrix tablet ($p > 0.05$). Therefore, sodium bicarbonate could be used to improve the wettability of THE formulation containing NRL as a binder without affecting the water sorption and erosion properties. SEM micrographs of all dried granules are shown in Figure 2. Wet granulation with NRL as binder provided good characterized dried granules similar to formulation with PVP-K30 as binder. They showed irregular shape of dried granule and good uniformity of granule size at magnification of 15X.

In vitro release profile of THE from tablets using NRL and PVP-K30 as the binders are shown in Figure3. PVP tablet showed higher drug release during the first 2 hrs than Non-Bicarb NR and Bicarb NR matrix tablets. Thus, the hydrophilic properties of PVP-K30 promoted THE release [3]; nevertheless, it showed steady state of drug release at about 90% whereas Non-Bicarb NR matrix tablet could control the drug release until 100% at 6 h. Therefore, NRL could be used as the binder in controlled

IOP Conf. Series: Materials Science and Engineering 1234 (2022) 012001 doi:10.1088/1757-899X/1234/1/012001

drug release solid dosage [4,5]. Surprisingly, Bicarb NR matrix tablet retarded the drug release and control the drug release up to 8 h. This result could be described from the mild alkaline property of sodium bicarbonate which, like the ammonia, could stabilize NRL. This remained alkaline environment promoted the easier distribution of NR in matrix component during mixing of wet granulation process; therefore, this obtained matrix efficiently modulated the drug release. After release test, Bicarb NR matrix tablet was dried then examined for the surface topography. There were many pores on the surface of Bicarb NR matrix tablet, indicating the channel generated from CO_2 gas bubbles (Figure 2). The linear regression of four different kinetic models were fitted with release profiles and their coefficients of determination (r^2) and model selection criterion (MSC) were calculated (data not shown). The first-order was found to be the best model for Non-Bicarb NR matrix tablet and PVP tablets; thus, the release behaviors were based on drug concentration dependence. Whereas Higuchi's equation was the best model for describing the THE release from Bicarb NR matrix tablet. The obtained exponent value (n) from power law expression was about 0.5 for Non-Bicarb NR and Bicarb NR matrix tablets, which indicated drug release governed by diffusion mechanism. These results were consistent with the previous reports [4,5]. In case of PVP tablet, n value was less than 0.5, signifying a coupling of drug diffusion, as well as polymer relaxation, so-called anomalous transport [3]. In addition, the allergenic protein from NRL has not been reported to affect the drug release behavior [5,6]. Nevertheless, the consideration about allergy from NRL protein should be concerned.

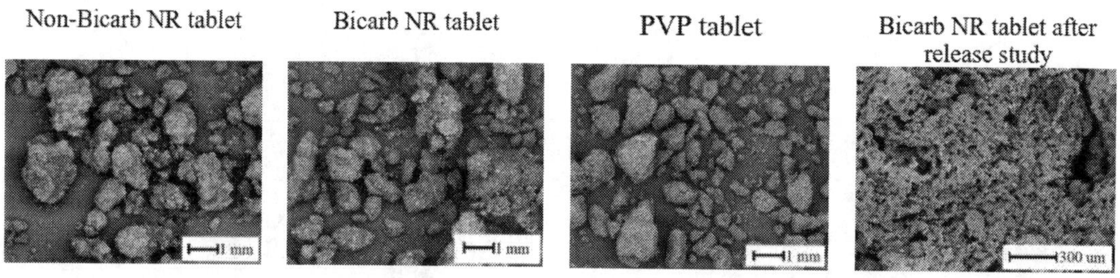

Figure 2. SEM micrographs of THE dried granules at magnification of 15X and Bicarb NR tablets after release study at magnification of 75x.

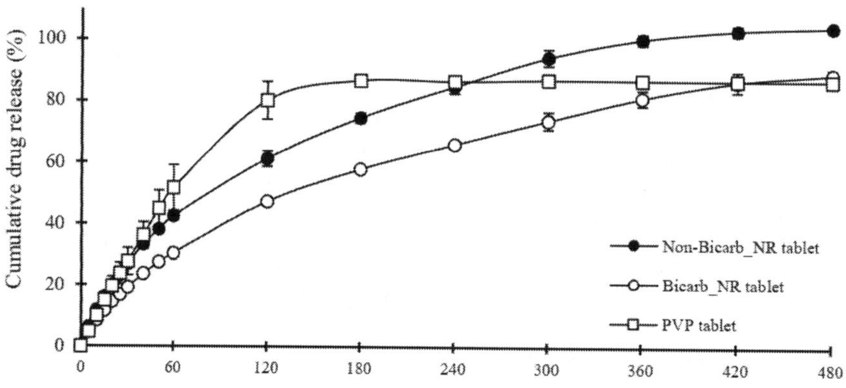

Figure 3. *In vitro* release profiles of THE monolithic matrix tablets fabricated with different binders in 0.1 N HCl buffer pH 1.2 (n=6).

4. Conclusions

THE monolithic matrix tablets comprising PVP-K30 and NRL as the binders in formulations were successfully fabricated with wet granulation and tableting processes. Non-Bicarb NR and Bicarb NR matrix tablets exhibited the similar physical properties to those of PVP tablet. In addition, the controlled

THE release with diffusion mechanism was presented in both formulations. Surprisingly, the addition of sodium bicarbonate in THE monolithic matrix tablets employing NRL as a binder could also prolong the drug release. Nevertheless, other polymers should be also used to compare with natural rubber latex as binder for extended-release tablet in further study.

Acknowledgement
The researchers are grateful to School of Pharmacy, Walailak University and the Research and Creative Fund, Faculty of Pharmacy, Silpakorn University, Thailand.

References

[1] Panrat K, Boonme P, Taweepreda W and Pichayakorn W 2012 *Procedia Chem.* **4** 322-27.
[2] Hansel TT, Tennant RC, Tan AJ, Higgins LA, Neighbour H, Erin EM and Barnes PJ 2004 *Drugs Today.* **40(1)** 55.
[3] Kasperek R, Zimmer L, Zun M, Dwornicka D, Wojciechowska K and Poleszak E 2016 *Curr. Issues Pharm. Med. Sci.* **29(2)** 71-8
[4] Panrat K, Boonme P, Taweepreda W and Pichayakorn W 2013 *Adv Mat Res.* **747** 91-4
[5] Herculano RD, Guimaraes SAC, Belmonte GC, et al 2010 *Mater. Res.* **13** 57-61
[6] Murbach HD, Ogawa GJ, Borges FA, et al 2014 *Int. J. Biomater.* **2014** 1-7.

IOP Conf. Series: Materials Science and Engineering 1234 (2022) 012002 doi:10.1088/1757-899X/1234/1/012002

Finite Element Analysis of elastomer: Case study – Rolling resistances of pneumatic and solid tyres

P Jittham*, S Sucharitpwatskul*, S Siriruk*, S Meesaringkarn**

*National Metal and Materials Technology Center (MTEC)
National Science and Technology Development Agency (NSTDA)
114 Thailand Science Park, Phahonyothin Rd., Klong Nuang, Klong Luang, Pathum Thani, 12120, Thailand
Phone: +662-564-6500, Fax: +662-564-6501
**Rubber Technology Research Centre (RTEC)
Faculty of Science, Mahidol University, Phuttamonthon 4 Rd., Salaya, Phuttamonthon, Nakorn Pathom 73170, Thailand

*Corresponding author: pairotej@mtec.or.th

Keywords: Finite Element Analysis, Rolling Resistance, Tread Design, Pneumatic Tyre, Solid Tyre

Abstract. EU tyre labelling rules have been applied since November 1st, 2012, to inform about the performances of tyres on fuel efficiency (rolling resistance), safety (braking on wet surfaces) and noise (external noise). The tyre labels help customers to make their purchase decisions by trade off on tyre performances and prices. Consequently, manufacturers have to engineer their products so that their tyres are classified on top of the label categories. Tyre rolling resistance involves in many kinds of tyre knowledge such as rubber formulations, tyre structures and tyre tread patterns. Generally, 70% of tyre rolling resistance depends on the hysteresis properties of rubber compounds and 30% depends on other properties such as road conditions, tyre pressures, load carrying, vehicle speeds, and tyre tread designs. From this point of view, even if rubber compounds are formulated to have good hysteresis properties, they can help reduce tyre rolling resistance up to 70%. If tyre rolling resistance needs to be further reduced, tread designs have to be considered. Therefore, this case study will show how to reduce tyre rolling resistance by using Finite Element Analysis (FEA). Pneumatic light truck tyres were used to study their tread patterns, tread depths and contact areas affecting their rolling resistances, while solid tyres were used to study on their tread patterns, tread depths, contact areas and tyre structures affecting their rolling resistances. The FEA results showed that tread patterns had high effects on rolling resistance of pneumatic light truck tyres while they had little effects on solid tyres. Contact areas and tread depths affected rolling resistances of both tyres. Increase in contact areas reduced tyre rolling resistances while increase in tread depths resulted in higher tyre rolling resistances.

1. Introduction

Whenever fuel oil price increases remarkedly, it impacts on living cost of many people. There are many ways to save fuel oil of vehicles such as avoid long idling, eliminating unnecessary weight, avoid aggressive driving, and using energy saving tyres. Tyre rolling resistance has an impact on fuel consumption and it is an indicator for energy efficiency of a tyre [1-3]. Hence, EU tyre labelling rules have been applied since November 1st, 2012, to inform about tyre performances, including fuel efficiency (rolling resistance) as shown in Figure 1. [4]. Rolling resistance of a tyre depends on hysteresis in a rubber compound approximately 70% and on other properties approximately 30% such as road conditions, tyre pressures, load carrying, vehicle speeds, and tyre tread designs [1-3]. Although, rubber compounds are formulated to have good hysteresis properties, they can help reduce rolling resistance up to 70%. To further reduce rolling resistance, other properties including tyre tread pattern need to be concerned. Finite Element Analysis (FEA) is a powerful tool used to design structures and tread patterns so that tyre rolling resistances can be foreseen [5-7]. Therefore, this case study will show how to reduce tyre rolling resistances by using Finite Element Analysis (FEA). Pneumatic light truck tyres were used to study their tread patterns, tread depths and contact areas affecting their rolling resistances, while solid tyres were used to study their tread patterns, tread depths, contact areas and tyre structures affecting their rolling resistances. The main advantages of FEA are time and cost reduction for product design process because interested behavior of the product can be seen previously without any prototypes. However, there are some disadvantages of FEA due to the need of FEA software and high-performance computers to solve the problems. These software and computers are costly. In addition, there are a lot of input data preparation such as geometry modelling, material property measurement and boundary condition specification. The accuracy of these input data is also the main source of uncertainty of FEA results.

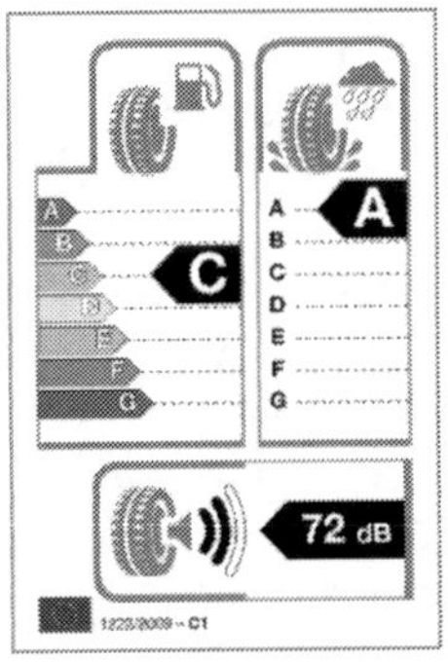

Figure 1. Example of tyre label.

2. Experimental

Pneumatic light truck tyres sized 195/65R14 and solid tyres sized 17.00 – 12 were used in this study. The tyre drum test machine Model dmt-350PC was used to measure tyre rolling resistance coefficients (RRCs) using a spindle method (Figure 2). The test conditions were shown in Table 1. FEA software (MSC.MARC-2012) was used to model and simulate this measurement as shown in Figure 3. The measuring and simulation results were compared for verification. Then, FEA was used to study effects of tread patterns, tread depths and contact areas on rolling resistance of pneumatic light truck tyres. Effects of tread patterns, tread depths, contact areas and tyre structures on rolling resistance of solid tyres were also studied.

Table 1. Test conditions for measuring RRCs at 38°C.

Tyre	Load (Kg)	Speed (km/hr.)	Duration (min)	Tyre pressure (psi)
Pneumatic light truck tyre	807.5	80	100	65
Solid tyre	2,000	10	120	-

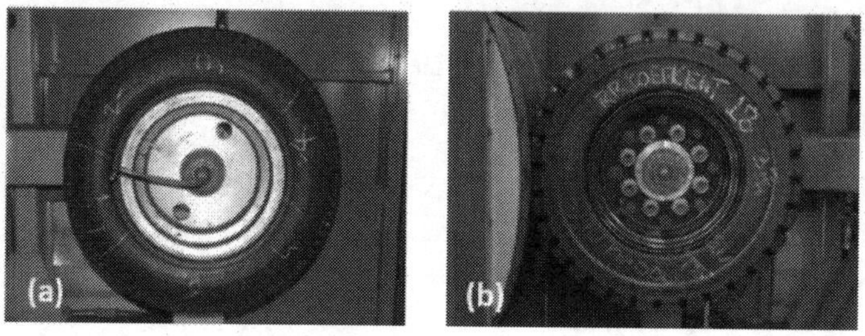

Figure 2. Tyre drum test machine for measuring RRCs of (a) pneumatic light truck tyres and (b) solid tyres.

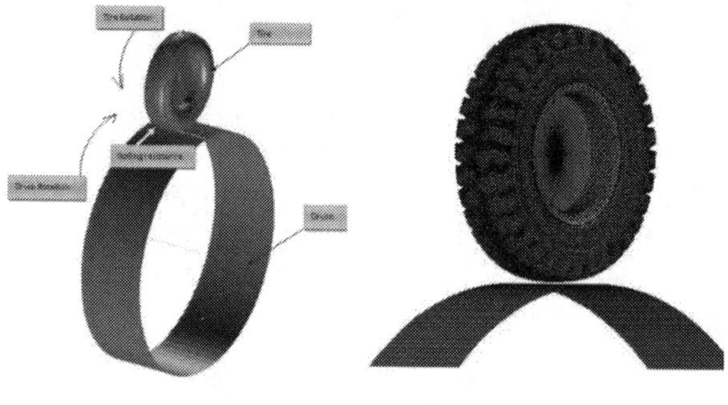

Figure 3. Simulation of RRCs measurement using FEA.

3. Results and discussion

The RRCs of light truck tyres and solid tyres measured by the tyre drum test machine and simulated by FEA are shown in Table 2. The effects of tread patterns, tread depths and contact areas on rolling resistance of light truck tyres are shown in Figure 4. The effects of tread patterns, tread depths, contact areas and tyre structures on rolling resistance of solid tyres are shown in Figure 5-6, respectively.

Table 2. RRCs measured by the tyre drum test machine and simulated by FEA.

	Rolling Resistance Coefficient (kg/ton)		% Difference
	Drum test	FEA	
Pneumatic light truck tyre	12.14±0.50	11.83	2.55
Solid tyre	31.02±1.75	27.08	12.7

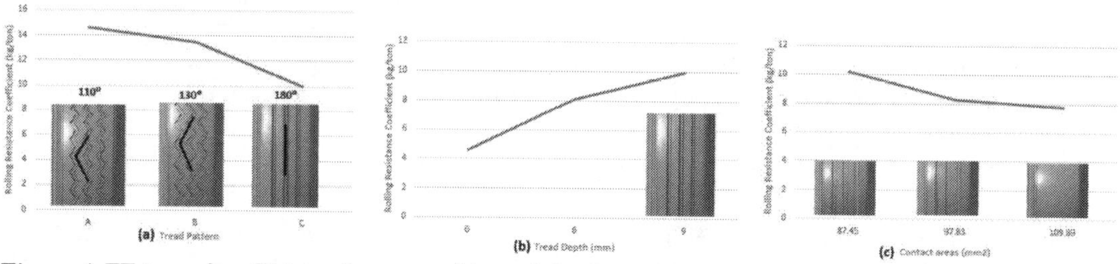

Figure 4. FEA results of (a) tread patterns, (b) tread depths and (c) contact areas vs RRCs of light truck tyres.

IOP Conf. Series: Materials Science and Engineering 1234 (2022) 012002 doi:10.1088/1757-899X/1234/1/012002

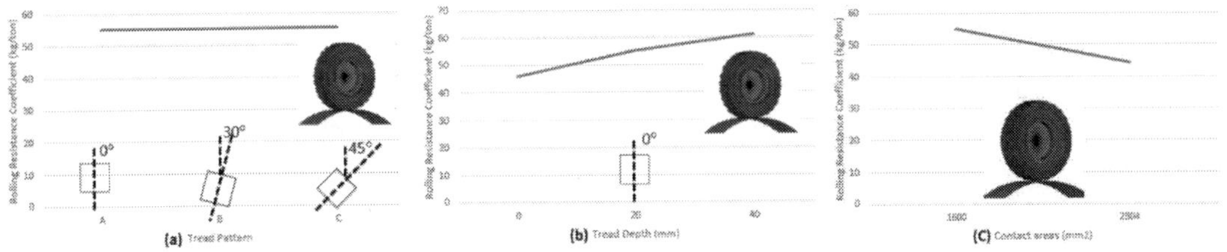

Figure 5. FEA results of (a) tread patterns, (b) tread depths and (c) contact areas vs RRCs of solid tyres.

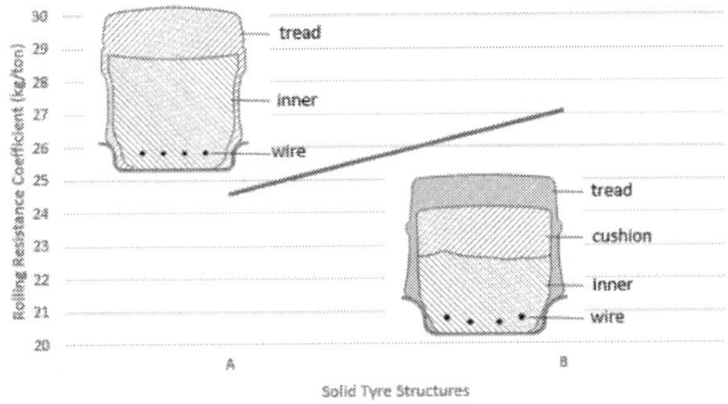

Figure 6. FEA results of tyre structures vs RRCs of solid tyres.

As showed in Table 1, RRCs of light truck tyres and solid tyres simulated by FEA were comparable with the results from the drum test machine with differences of 2.55% and 12.70%, respectively. This indicated that FEA was a useful tool to simulate RRCs getting from the drum test method. Tread patterns had high effects on RRCs of pneumatic light truck tyres (Figure 4 (a)) while they had little effects on solid tyres (Fig. 5 (a)). At higher testing speeds, RRCs increased due to the increase in work done in deforming of the tyres [8]. Therefore, under this condition, tread patterns were significant to RRCs of light truck tyres. Tread depths affected RRCs of both tyres. Increase in tread depths caused higher RRCs (Figure 4(b) and 5(b). As the tread blocks with higher depths were easily deformed and RRCs increased with increasing deformation [1-3], the higher tread depths showed higher RRCs than those of the lower tread depths. Increase in contact areas resulted in lower RRCs in both tyres (Figure 4(c) and 5(c) due to reduction of contact pressures of the contact area. In the case of tyre structures, 2-layered solid tyres had lower RRCs than those of 3-layered solid tyres (Figure 6). Due to the cushion layer, which made of a rubber compound softer than the rubber compounds in tread and inner layers, 3-layered solid tyres had more deformation under the same load and resulted in higher RRCs.

4. Conclusions

FEA is a useful tool to simulate RRCs of tyres without the need of manufacturing new tyres for RRCs measurement. Due to large deformation at a higher speed, tread patterns have high effects on RRCs of pneumatic light truck tyres while they have little effects on RRCs of solid tyres. Increase in tread depths will result in higher RRCs both in pneumatic light truck tyres and solid tyres because of easy deformation of higher depth tyre treads. Increase in contact areas will decrease RRCs because of the reduction of contact pressures in the contact areas. Finally, 2-layered solid tyres have lower RRCs than those of 3-layered solid tyres because 3-layerd solid tyres have cushion layers which result in more tyre deformation.

Acknowledgement

This project was funded by Thailand Science Research and Innovation (TSRI). The equipment and workspaces were supported by Rubber Technology Research Centre (RTEC) and National Metal and Materials Technology Center (MTEC). The authors would like to gratefully thank all of them.

References

[1] A. N. Gent and J. D. Walter, "The Pneumatic Tire", The University of Akron, 2005.
[2] SAE Society of Automotive Engineers, Automotive Handbook, 3rd Edition, USA, 1993.
[3] Bridgestone Americas Tire Operations, Real Answers: Tires and Truck Fuel Economy, Vol. 15 issue 1, 2010.
[4] Regulation (EU) 2020/740 of the European Parliament and of the Council on the Labelling of Tyres with Respect to Fuel Efficiency and Other Parameters, Amending Regulation (EU) 2017/1369 and Repealing Regulation (EC) No 1222/2009, 25 May 2021
[5] Boast, D. and Coveney, V. A., Finite Element Analysis of Elastomers, Antony Rowe Limited, Chippenham, Wiltshire, I999.
[6] Miller, K., Testing Elastomers for Hyperelastic Material Models in Finite Element Analysis, Testing and Analysis, Axel Products, Inc., Vol. 3, 2004.
[7] Reddy, J. N., An Introduction to Nonlinear Finite Element Analysis, Oxford University Press, New York. 2006.
[8] Heisler, H., Advanced Vehicle Technology, 2nd Edition, College of Northwest London, Willesden Centre, London, UK, 2002

IOP Conf. Series: Materials Science and Engineering 1234 (2022) 012003 doi:10.1088/1757-899X/1234/1/012003

Prediction of crosslink density of prevulcanised latex using NIR Spectroscopy based on combination of fractional order derivative (FOD) and variable selection methods

C H Lim[1], J Posom[2,*], P Sirisomboon[3]

[1]Thai Rubber Latex Group Public Company Limited, Nongyai, Chonburi, 20190, Thailand
[2]Department of Agricultural Engineering, Faculty of Engineering, Khon Kaen University, KhonKaen, 40002, Thailand
[3]Department of Agricultural Engineering, School of Engineering, King Mongkut's Institute of Technology Ladkrabang, Bangkok, 10520, Thailand

* Corresponding author: Jetspo@kku.ac.th

Abstract. Rapid method in measurement of crosslink density is required in factory. The objective of this study was to develop the prediction model of crosslink densities based on near infrared (NIR) spectroscopy method. The prediction models were developed using partial least squares regression (PLSR) with spectral pre-treatment of fractional order derivatives (FOD) and variable selection methods including successive project algorithm (SPA) and genetic algorithm (GA). The result demonstrated that prevulcanised (PV) latex model had higher accuracy than that of PV_{50} latex model. Effective model in predicting crosslink densities of PV and PV_{50} latices could be pre-processed with FOD=1 and 0.75, respectively. The prediction model generated with full wavelength had the standard error of cross validation (SECV) of 3.21% and 3.52%, respectively. The model performance of PV latex could improve with variable selection method of GA which reduced the SECV from 3.21% to 3.17% and number of wavelengths reduced from 1059 to 937. The model performance of PV_{50} could not reduce by using the variable selection method. However, the GA could reduce the number of wavelengths from 1059 to 216.

1. Introduction

Some of the key considerations and the way crosslink densities are influenced by vulcanization ingredients are well known. In practice, for optimal level, the crosslink density must be sufficiently high to prevent failure by viscous flow but not too high to avoid brittle fracture. Various types of crosslinks affect the chemical and mechanical properties of rubber namely covalence crosslink (polysulfidic, mono- and di-sulfidic and carbon-carbon bonds) and ionic crosslink.

NIR spectroscopy would be a suitable method for crosslink density determination because no chemical is required and also cheap to operate. In order to improve the NIR model, the factors of spectral pre-treatment and spectral variable selection method is important and could be studied. The task of spectral variable selection method helps to reduce overfitting by eliminating redundant features [1]. The use of derivatives for spectrum pre-processing has two major advantages [2]. First, the baseline zero is clearly defined which means that ratios of spectral values are not affected by variations in offsets that typically are found in absorbance spectra of diffusely scattering samples. Second, each higher derivative has more bands, and they are narrower than the original absorbance bands which make it easier for the regression

IOP Conf. Series: Materials Science and Engineering 1234 (2022) 012003 doi:10.1088/1757-899X/1234/1/012003

to find and utilise bands that are less influenced by interfering absorbers in the spectra. Considering that the absolute value of the derivative of the third or more is small and the spectral information is low, it is generally not recommended to use the third derivative or more [3]. Hong *et al.* [4] reported that fractional order derivative (FOD) can provide additional detailed spectral variation information for prediction of soil organic matter (SOM) compared with the original reflectance (order = 0). Commonly used integer order reflectance, that is 1- and 2-orders memory-based learning (MBL) outperformed the two global modelling approaches (PLS and RF). In almost all of FOD transformations except for 2-orders reflectance, the MBL yielded the lowest $RMSE_{val}$ values (ranging from 4.36 to 5.55) as well as the highest RPD values (ranging from 1.75 to 2.23). The best validation result was produced by 0.25-order reflectance ($R^2_{val} = 0.82$) and followed by 0.5-order reflectance ($R^2_{val} = 0.81$) and their predictive accuracies were all classified as very good predictions [4]. Furthermore, the fitted line between measured and estimated SOMs and the scatter points for MBL model based on 0.25-order reflectance were closer to 1:1 line than other MBL models. The fourth derivative spectra analysis has been applied to the UV-VIS spectrophotometry. Faust and Norris [5] noted that the use of the 4th derivative provided additional sharpening of absorption bands, as well as accentuating weakly absorbing components. Bidigare *et al.* [6] indicated in analysis of natural particulates collected from the Sargasso Sea, a dynamic partitioning algorithm based on the 4th derivative spectrum can solve these problems because it has the benefits of sharpening the peaks and valleys of a spectrum without changing their positions. This research was an attempt to examine the optimal FOD pre-treatment on PV latex spectra and suitable variable selection method for evaluation of the crosslink densities using PLS regression. Sub objective included (1) to find the optimal FOD, (2) to find the best variable selection method and (3) to find the best model development method.

2. Methodology

2.1. Samples

The PV and PV_{50} natural rubber latex samples for the experiment were collected from factory of Thai Rubber Latex Group Public Company Limited in the Nongyai district, Chonburi province, Thailand. Each week, three samples of PV latex were drawn from each tank at different levels (upper, middle, and lower) for eight consecutive weeks starting from the date of production (totally nine times). The PV_{50} latex was prepared from the PV latices of all the batches diluted to 50% solids content, hence PV_{50}. In total, 157 latex samples with different crosslink densities were obtained.

2.2. NIR spectra acquisition and reference method

The Vis -NIR spectrometer (AvaSpec-2048-USB, Netherlands) with the wavelength range of 350 and 1100 nm, was applied. Before scanning, the Spectralon material was scanned to get white reference. PV spectra and PV_{50} latices were scanned by immersing the probe into each of the sample in the holder cups. The scanning method of reflectance mode with integration times of 1.20 ms was set. Each sample was scanned for three time and then the average calculated. After that the TS determination was carried out on the latex samples. The toluene swell method was conducted for measuring the crosslink density of PV latex as reported by Lim and Sirisomboon [7]. The percentage toluene swell was calculated as follows: $\% \text{ swell} = \frac{Y-X}{X} \times 100$, where X is initial diameter and Y is swollen diameter read from the graph paper.

2.3. PLS modelling

After collecting NIR spectra and TS, the model development was done. PLS regression combined with different FOD and spectra variables selection (i.e. successive project algorithm (SPA) and genetic algorithm (GA)) were used for model development. Figure 1 shows flow diagram for model development. PLS model was performed by MATLAB R2021a (Mathworks Inc., USA).

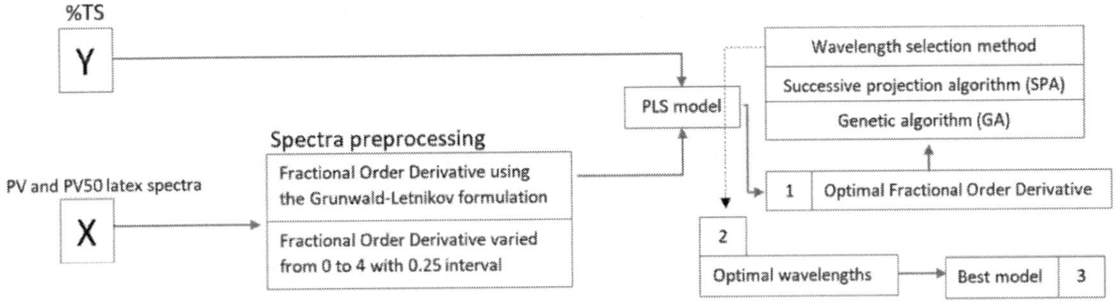

Figure 1. Schematic diagram for model development.

The FOD of 0, 0.25, 0.5, +++, and 4 were used, where order = 0 was raw absorbance spectra. FOD method was used for signal filtering and pattern recognition [4] where FOD algorithm was used to compute the fractional derivative of a sampled function using Grunwald-Letnikov formulation [4]. The data analysis had three steps: (1) PLS models were generated from different FOD, (2) the best FOD was selected (any FOD giving the lowest standard deviation of cross-validation (SECV) was selected). The selected FOD spectra was extracted for the important wavelength using SPA and GA (variables were sorted in descending order from the highest important wavelength to the lowest wavelength) and (3) their performances were compared. Any model providing lowest SECV was selected. Coefficient of determination (R^2), SECV, bias and ratio of prediction to deviation (RPD) were determined.

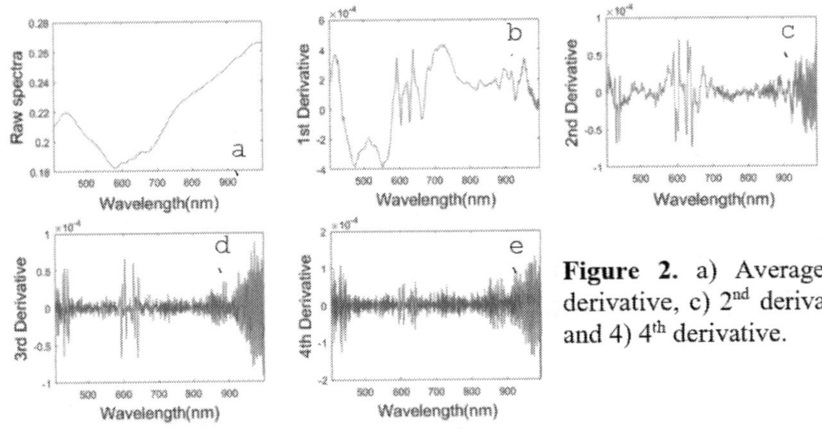

Figure 2. a) Average raw spectra, b) 1st derivative, c) 2nd derivative, d) 3rd derivative, and 4) 4th derivative.

3. Result and discussion

3.1 NIR spectra and reference value

Figure 2 shows the FOD of PV latex including raw spectra, 1st derivative, 2nd derivative, 3rd derivative, and 4th derivative. The wavebands were narrow where FOD increased. The statistical data of toluene swell gave the maximum, minimum, average, and standard deviation (SD) of 118, 82, 91.6, and 7.3%, respectively.

3.2 Model development result

Table 1. PLS regression results of PV and PV_{50} latices developed from different FOD and spectral variable selection method.

Latex	Wavelength selection	Number of wavelengths	FOD	PLS factor	R^2_{cv}	SECV	RPD
PV	NON	1059	1	7	0.81	3.21	2.3
	SPA	1049	1	7	0.81	3.19	2.3
	GA	937	1	7	0.81	3.17	2.3
PV_{50}	NON	1059	0.75	7	0.78	3.52	2.1
	SPA	1025	0.75	7	0.78	3.52	2.1
	GA	216	0.75	7	0.75	3.68	2.0

The result of PLS regression was demonstrated in Table 1 which showed the wavelength selection method, number of wavelengths, FOD, PLS factor, R^2_{cv}, SECV and RPD. The effective model of PV latex prediction was found when developed with the 1st FOD which provided PLS factor, SECV and R^2_{cv} of 7, 3.21 and 0.81 (see Figure 3a and b), respectively. Meanwhile, the PV_{50} latex model developed using 4th FOD had highest accuracy, gave the lowest SECV and highest R^2_{cv} of 3.16 and 0.81, respectively (see Figure 3c and d). However, the effective model of PV_{50} latex could be developed with FOD of 0.75 because SECV started to be constant, gave SECV and R^2_{cv} of 3.52 and 0.78, respectively (see Table 1 & Figure 3c and d). Therefore, the suitable FOD for PV and PV_{50} were 1 and 0.75, respectively. After that NIR spectra pretreated with 1 and 0.75 were taken to the SPA and GA algorithms to extract the important variables with the purpose for improving model performance.

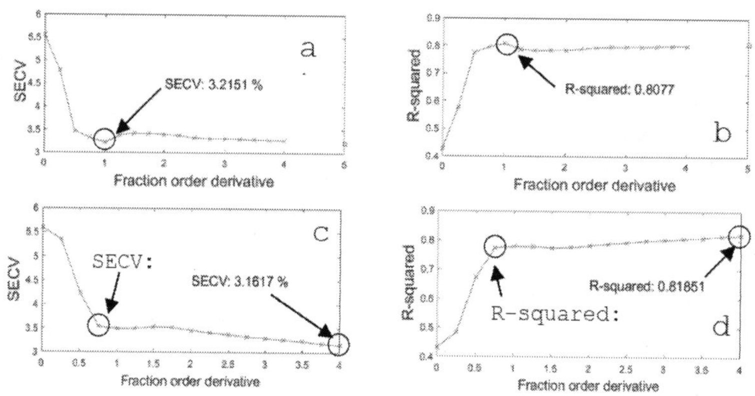

Figure 3. PLS regression result validated using one full cross validation obtained from different FOD of PV latex, a) SECV of PV, b) R^2 of PV, c) SECV of PV_{50}, and d) R^2 of PV_{50}.

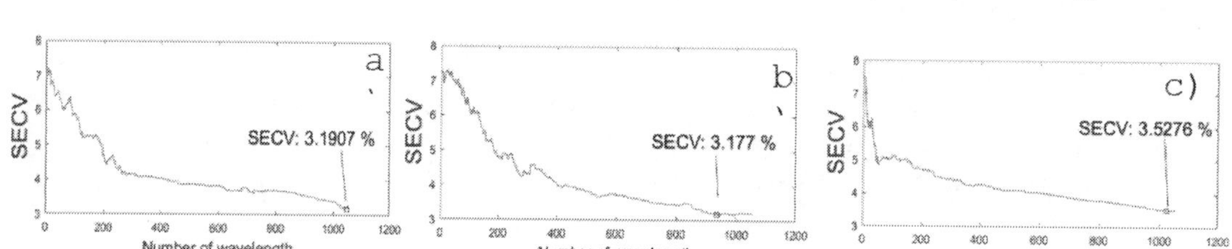

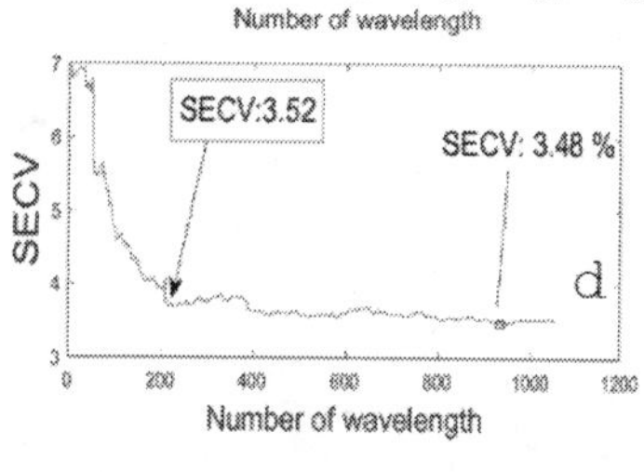

Figure 4. PLS regression result validated using one full cross validation of FOD=1 of PV latices, (a) PV latex FOD=1 extracted by SPA, (b) PV latex FOD=1 extracted by GA, (c) PV_{50} latex FOD=0.75 extracted by SPA and (d) PV_{50} latex FOD=0.75 extracted by GA.

Figure 4. shows the PLS regression result of PV and PV_{50} latex models generated using FOD=1 and 0.75 combined with SPA and GA. For the PV latex model, combining SPA can increase accuracy slightly (SECV reduced from 3.21 to 3.19%). Combining GA can further improve accuracy (SECV reduced from 3.21 to 3.17%). Meanwhile, number of variables can reduce from 1059 to 937. For PV_{50} latex model, combining SPA cannot increase accuracy but reduce the number of variables from 1059 to 1025, but SECV remain the same. Combining GA reduced the number of variables from 1059 to 216, but the SECV increased slightly from 3.52 to 3.68%.

4. Conclusions

Understanding the crosslink density change is necessary in production process. NIR spectroscopy is a rapid and cheap method. The prediction model was developed with different FOD and variable selection method. The result showed that PV latex model had slightly higher performance than that of PV_{50} latex model. The FOD=1 and 0.75 were suitable for PV and PV_{50} latices and wavelength selection of GA was optimal to improve model accuracy. The model performance of PV latex can increase (SECV reduces from 3.21 to 3.17% and number of wavelength reduces from 1059 to 937). The model performance of PV_{50} latex did not increase but number of wavelength reduces from 1059 to 216. This experiment can be improved by adding samples covering the full range of crosslink densities (i.e., 75 to 160% TS).

Acknowledgments

The authors would like to express their sincere thanks to Thailand Research Fund for the financial support (Grant number RDG5950011).

Reference

[1] Lee G, Lee K 2021 *BioData Mining* **14** 7.
[2] Hopkins DW 2016 *NIR news* **27(7)** 23-28.
[3] Pan T, Zhang J, Shi X 2019 *NIR news* **31(1-2)** 24-27.
[4] Hong Y, Chen Y, Yu L, Liu Y, Liu Y, Zhang Y, Liu Y, Cheng H 2018 *Remote Sens.* **10** 479.
[5] Faust MA, KH Norris 1985 *Limnol. Oceanogr* **30** 1316-1322.
[6] Bidigare RR, Morrow JH, Kiefer DA 1989 *J. Mar. Res* **47** 323-341.
[7] Lim CH, Sirisomboon P 2018 *J Near Infrared Spec* **26(3)** 159-168.

Study of porous rubber pipes reinforced with waste tire fiber and pineapple leaf fiber for smart irrigation system

M Junpunya[1], B Ou-udomying[2], T Amornsakchai[3] and I Jangchud[1,*]

[1] Department of Chemistry, School of Science, King Mongkut's Institute of Technology Ladkrabang (KMITL), Bangkok 10520, Thailand.
[2] Saeng Thai Rubber Co., Ltd., Samutprakan 10130, Thailand.
[3] Department of Chemistry and Center of Excellence for Innovation in Chemistry, Faculty of Science, Mahidol University, Nakhon Pathom 73170, Thailand

*Corresponding author: *ittipol.ja@kmitl.ac.th*

Abstract. Global trend in waste management involves sustainable development for waste utilization according to the concept of circular economy. Wastes from worn-out tires have been known as a world problem and not yet to be resolved successfully. In this work, wastes from worn-out tires, i.e., ground tire rubber (GTR), reclaimed rubber (RR) and waste tire fiber (WTF) were used to produce reinforced porous rubber pipes as irrigation system for smart farming. Comparative study of porous rubber pipes reinforced with natural fibers (pineapple leaf fiber, PALF) was also carried out. The experiment was divided into 2 parts, i.e., porous sheets (model study) and extruded porous pipes. Compound samples were prepared by mixing GTR, RR, and natural rubber (NR) at a ratio of 60:20:20 (by weight) with other additives by using an internal mixer, then shaped into samples by using a compression molding machine (sheets) and a single-screw extruder (pipes). Samples were vulcanized by a hot air oven. Factors affecting sample properties were studied, such as, fiber types (WTF and PALF) and fiber loadings (0-15 phr). Properties of the samples were characterized, such as, mechanical properties, morphology, water permeability rate, and %diameter swell. It was found that tensile strength was increased up to an optimum point. Modulus (at 10% strain, M10) and hardness were increased with increasing fiber loading. However, %elongation at break had a trend to decrease. Morphologies of the composites made from both fibers were insignificantly different in fiber distribution. By adding the fibers, %diameter swell was clearly improved. Pipe samples were also tested in precision smart agriculture system. It can be proven that WTF and PALF can be used to reinforce the porous pipes and used as a watering device for precision irrigation system.

1. Introduction

Worn-out tires cannot decompose easily as they are complex objects consisting of many components such as crosslinked rubbers, steel wires, fibers, and other additives [1]. Accumulation of waste tires leads to a serious waste management issue and a persistent environmental pollution. To build a sustainable economy, several waste management strategies have been proposed to deal with the increasing number of worn-out tires including retreading, reclaimed rubbers, ground rubbers, etc.[2] WTF is classified as a unique waste according to the European Waste Code 19.12.08 since they have to be burned and/or buried. Nevertheless, these activities pose additional threats, including toxic hazards to soil pollution and groundwater contamination [3-4].

Natural fibers are sustainable materials which are easily available in nature and have advantages like low-cost, lightweight, renewability, biodegradability, and high specific properties. In Thailand, natural fibers are abundantly available both naturally and from agricultural wastes, such as jute, sisal, pineapple leaf, coir, bamboo, rice husk, vetiver grass, etc. The one that is abundantly available and is used very little is PALF. It has good mechanical properties and can be used to reinforce rubbers yielding high modulus and strength [5].

Porous pipes are effective watering devices due to low water consumption, less water evaporation and water consistently and efficiently to plant roots [6]. In previous works [7-8], porous pipes made from rubber were unsuccessfully done due to high %diameter swell and non-linear water rate at high pressure that it was difficult to control. In this work, reinforced porous pipes were made form GTR, RR, NR and fibers. The objective of this work was to study comparatively the effects of WTF and PALF fiber loading to the rubber formulas on the pipe properties resulting in optimum fiber loadings with good mechanical and physical properties. It can be used as a watering device in precision irrigation for smart farming systems.

2. Experimental

2.1 Materials and chemicals

STR 5L NR was purchased from Yong Seng Enterprise Part., Ltd., Thailand. RR, 30 mesh GTR and WTF were obtained from Saeng Thai Rubber Co., Ltd., Thailand. UCD-105 grade RR has 50% rubber content, 29% carbon black content, and Mooney viscosity (ML1+4 (100°C)) of 52. WTF with average diameter of 0.027 mm and 3-4 mm in length was used in this work. PALF with average diameter of 0.063 mm and 6 mm in length was obtained from Mahidol university. Details of PALF preparation and properties were described elsewhere [9]. N-cyclohexyl-2- benzothiazyl sulfenamide (CBS), sulfur, zinc oxide (ZnO) and stearic acid were used for the curing system. A blowing agent and kicker were used to create open-cell structures. All chemicals were used as received. Formulae of rubber compounds were kept proprietary.

2.2 Preparation of 40% fiber masterbatches

NR was masticated in an internal mixer (CT, MX500-D75L90) for 1 min with a rotating speed of 40 rpm and 70°C.Then, fiber (WTF, PALF) was gradually added during 3 min loading time. Fiber masterbatch was then rolled into sheet on a two-roll mill (LabTech, LRM200). Sheeted rubber compound was conditioned at room temperature for 24 h before measurements taken.

2.3 Preparation of porous rubber sheets and extruded porous rubber pipes

Rubbers, GTR, fiber and all chemicals were mixed by using an internal mixer with a rotating speed of 40 rpm and 70°C. The weight ratio of GTR, RR and NR at a 60:20:20. Fiber loadings (WTF and PALF) were varied as 0, 3, 6, 9, 12 and 15 phr. Total mixing time was kept constant at 17 min. Samples were shaped into samples by using a compression molding machine (sheets) (LabTech, LP20) with size 150 x150 x 4 mm. The upper plate was lifted shortly for 5 s. A single-screw extruder with a screw speed of 30 rpm at 60°C and die diameters of 14 mm (inside) and 18 mm (outside) was used to prepare rubber pipe samples. Pipe samples were vulcanized in a hot air oven at 160°C for 20 min.

2.4 Characterization and testing

Tensile properties were also examined using a universal testing machine (Cometech, QC-536M1) with a crosshead speed of 500 mm/min and the gauge length of 25 mm. Mechanical properties were measured according to ASTM D412. Morphology of porous samples were observed with a scanning electron microscope (SEM) (LEO, LEO 1455 VP). All samples were coated with gold prior to examination. Water permeability and %diameter swell of porous rubber pipe with changing water pressure were determined according to ASTM D3767-03.

3. Results and discussion

Figure 1(a) shows porous rubber pipe made from GTR, RR and NR. It can be used in the smart farming (irrigating) system by connecting the pipe with smart valve and pump systems (Figure 1(b)). Figure 2(a) displays tensile strength (solid line) and %elongation at break (dash line) of samples with fiber types and fiber loadings. Surprisingly, tensile strength of both fibers were insignificantly different and a trend to increase to an optimum fiber loading as fiber loading was increased. However, at high fiber loadings, %elongation at break of PALF was significantly decreased. This might be derived from defects of fiber agglomeration resulted from poor fiber dispersion difficulty in rubber mixing with high fiber loadings. This finding was confirmed by SEM images (3a) and watering tests as high water leaking out with visible holes in the samples. Figure 2(b) shows the effect of fiber loading on M10 (solid line) and hardness (Shore A) (dash line) of composites. Both M10 and hardness seem to rise significantly with increasing fiber loading. The observed behavior is in agreement with other fiber-reinforced rubber systems [10]. In addition, PALF has slightly higher M10 and hardness than WTF since PALF is lignocellulosic material with higher modulus. Typical modulus of the PALF fiber is 20 GPa [11] compared to 2-3 GPa for WTF fiber [12].

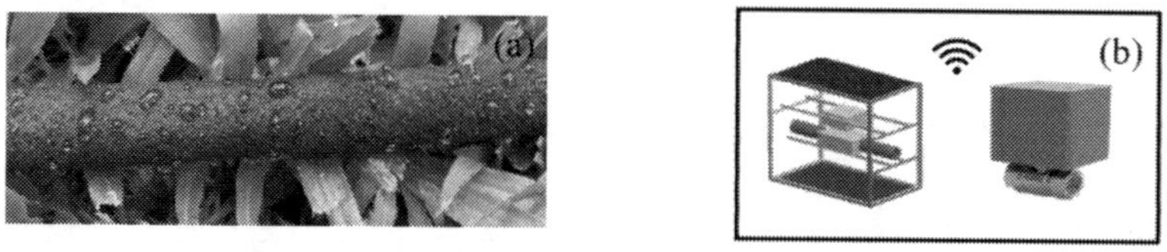

Figure 1. (a) Porous rubber pipe and (b) Smart pump and valve systems.

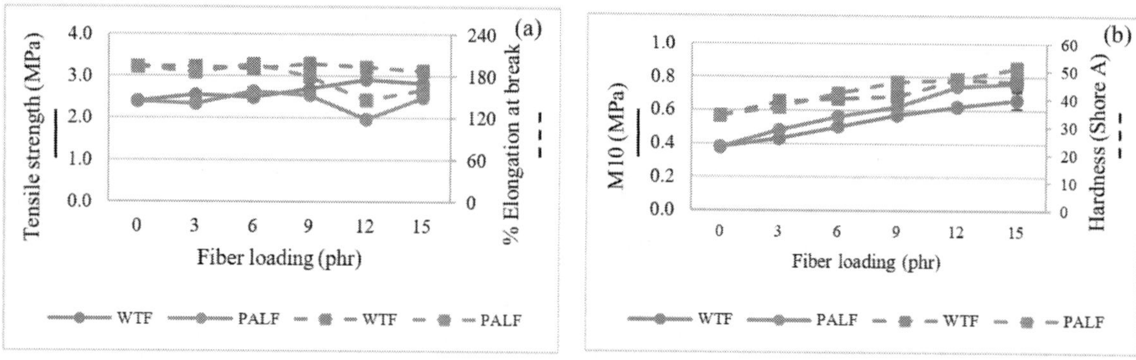

Figure 2. Effects of fiber loading on (a) tensile strength and %elongation at break (b) M10 and hardness of porous rubber samples.

Figure 3(b-c) displays SEM micrographs of cryogenic-cracked samples with 6 phr PALF and WTF. It was revealed that PALF and WTF were distributed well in rubber matrix. Voids distribution in the samples were good, but unequal in shape and size.

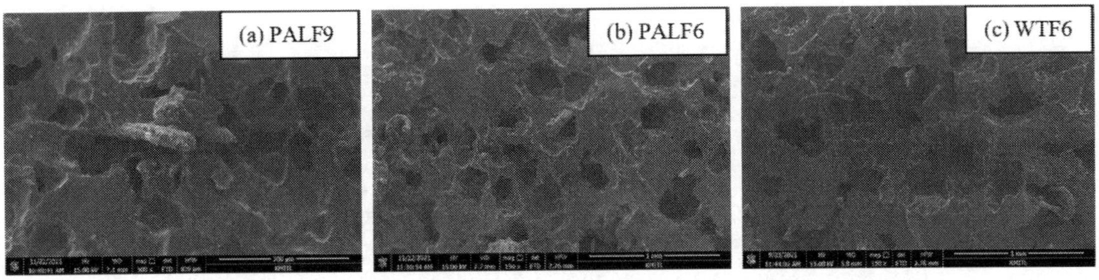

Figure 3. SEM micrographs of cryogenic-cracked samples (a) 500x of PALF 9 phr, (b) 150x of PALF 6 phr and (c) 150x of WTF 6 phr.

%Diameter swells of porous pipes as affected by water pressure were shown in Figure 4(a). Effect of fiber reinforcement was clearly observed as %swell significantly decreased when PALF or WTF loading was increased in the compounds. %Diameter swell of porous pipe can be related to water permeability as shown in Figure 4(b). As expected, water permeability rate was increased as water pressure was increased. More linear water permeation was observed when water pressure was increased. However, PALF has linear water permeation higher than WTF.

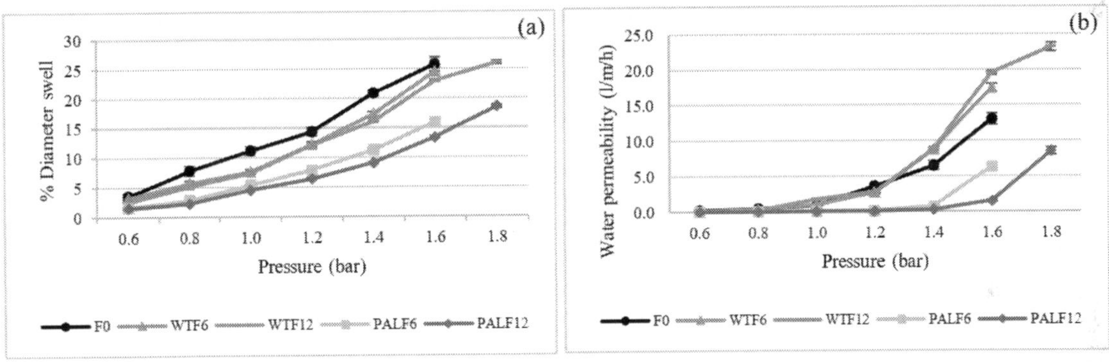

Figure 4. (a) %Diameter swell and (b) water permeability of porous pipes affected by water pressure.

4. Conclusions

Comparative study of porous rubber pipes made from worn-out tire wastes reinforced by waste tire and natural fibers were carried out in order to be used as irrigating device in precision and smart farming systems. It was found that mechanical properties had a trend to increase to an optimum fiber loading as fiber loading was increased. However, PALF has slightly higher modulus than WTF. By adding the WTF or PALF, %diameter swell was clearly improved. In this work, it can be proven that WTF and PALF can be used to reinforce the porous pipes and used as a watering device for precision irrigation systems.

Acknowledgement

Authors would like to thank King Mongkut's Institute of Technology Ladkrabang (KMITL) and the National Science and Technology Development Agency (NSTDA) for financial support throughout this work.

References

[1] Luo K, Wei Z, Xiuying Z, Xiujuan W and Sizhu W 2018 *Mater. Des.* **154** 312-325.
[2] Tseng H, Lin Z, Chen S, Lai W and Wey M 2019 *J. Clean. Prod.* **237** 117739.

[3] Yilmaz Y 2015 *Eng. Geol.* **188** 168-177.
[4] Thai Q.B, Chong R.O, Nguyen P.T.T, Le D.K, Le P.K, Phan-Thien Nand Duong H.M. 2020 *J.Environ. Chem. Eng.* **49** 104279.
[5] Wisittanawat U, Thanawan S and Amornsakchai T 2014 *Polym. Test.* **35** 20-27.
[6] https://homeguides.sfgate.com/porous-pipe-irrigation-vegetable-garden-54367.html: September 2021.
[7] Jammuan S 2004 Production of Porous Pipe made from Natural Rubber and Ground Rubber tire., M.Sc. Thesis, KMITL, Thailand.
[8] Pinitkul P 2018 Study of porous pipe made form GRT mixed NR and reclaimed rubber for agricultural applications., M.Sc. Thesis, KMITL, Thailand.
[9] Kengkhetkit N and Amornsakchai T 2012 *Ind. Crops Prod.* **40** 55-61.
[10] Agarwal K, Setua D.K, Mathur and G.N 2002 *Sci. J.* **52** 337-346.
[11] Surajarusarn B, Traiperm P and Amornsakchai T 2019 *Sains Malays.* **48** 145-154.
[12] Tian L, Wang D and Wei Q 2019 *J. Eng. Fiber Fabr.* **14** 1-7.

Study of bladder release agent formulas and methods to evaluate their efficiencies by using reciprocating tribometer

N Tanboriphan[1], B Teeraprawatekul[2], S Mahathanabodee[3,4], and I Jangchud[1*]

[1]Department of Chemistry, School of Science, King Mongkut's Institute of Technology Ladkrabang (KMITL), Bangkok 10520, Thailand
[2]Acme International (Thailand) Limited., Bangkok, 10250, Thailand
[3]Department of Production and Robotics Engineering, Faculty of Engineering, King Mongkut's University of Technology North Bangkok (KMUTNB), Bangkok 10800, Thailand
[4]Material Manufacturing and Surface Engineering Research Center (MaSE), Faculty of Engineering, King Mongkut's University of Technology North Bangkok (KMUTNB), Bangkok 10800, Thailand

*Corresponding author: *ittipol.ja@kmitl.ac.th*

Abstract. Bladder is one of the most important components in tire manufacturing process. In tire industry, methods to prolong bladder service-life and good bladder coatings for non-stick surface has been acquired. Bladder coated with good release agents can help reducing nonconforming products and increasing production efficiency. In this work, semi-permanent release agent formulas were studied. A new method to evaluate release agent efficiency by using a reciprocating tribometer were also carried out. Polydimethyl siloxane (PDMS) release agent formulas were developed with and without polymethyl hydrogen siloxane (PMHS). Effects of metal salt cure catalyst were also studied. In this work, method of release efficiency test was developed by using a reciprocating tribometer according to ASTM G133-05. It can be used to determine coefficient of friction of the coated surface and durability of the coatings. Coating durability test was also confirmed by repeating compression molding test to determine the number of release cycles. It was found that film durability was improved by adding the metal salt catalyst and PMHS. As the PMHS loading was increased, film endurance was improved to an optimum loading. Friction coefficient of the coatings could be related to the mold release ability and durability of the release agents. The technique can be used to estimate the service-life of the coated bladders. It can be used as a guideline for development of mold release agents in tire manufacturing industry.

1. Introduction

Release agents are chemicals used to prevent other materials from bonding to surfaces. They have been used in many industries such as medium density fiber board (MDF), fiber-reinforced composites, tire industry, etc. which require different release agents depending on their processes.

In tire manufacturing process, pneumatic tires are produced by molding and curing a green (or uncured) tire in a molding press. The green tire is pressed against a mold surface by means of an inner expandable "bladder". Therefore, a release agent is needed to apply on the bladder surface in order to prevent adhesion between the bladder and inner liner of green tire in hot press of vulcanization process. For a semi-permanent release agent, its efficiency can be determined by numbers of cycles before re-applying the agent in the vulcanization process. Good bladder release agents can reduce nonconforming (NC) products, increase production speed, and extend bladder service-life.

In previous work [1-2], PDMS and PMHS were used as release agents. They provided excellent slippage yielding longer release ability [3]. The water-based release agent had a lower number of cycles carried out by the hot compression method. And the film endurance test was not carried out by the tribometer. In this work, the PDMS and PMHS have much better release ability and were evaluated both techniques. The objective of this work was to develop PDMS bladder release agents modified with PMHS. Effects of PMHS loading (0-6%) and addition of metal salt cure catalyst were studied. Another objective was to develop methods of release efficiency tests by using a reciprocating tribometer according to ASTM G133-05. It can be used to determine coefficient of friction and wear resistance of the coatings. Coating durability test was also confirmed by repeating compression molding test to determine the number of release cycles.

2. Experimental

2.1 Materials

Butyl rubber compound was purchased from Siamnavakam Co. Ltd., Thailand. PDMS, wax-based, and silicone-based release agents supported by Acme International Ltd., Thailand. PMHS and cure catalyst was purchased from Momentive Performance Materials Inc., USA. A wetting agent was obtained from BYK-Chemie Inc., Germany. Hexane was purchased from Koventure Co.Ltd., Thailand. All chemicals were used as received.

2.2 Sample preparation

Formulas of release agents are presented in Table 1. Some details of chemicals are kept proprietary. PDMS, PMHS, cure catalyst and wetting agent were mixed into hexane in 250 cm^3 beaker. The mixtures were then stirred at 250 rpm for 1 min by using a mechanical stirrer (IKA RW20). After mixing, each sample was placed in a closed container.

Table 1. Formulas of release agents in wt.%.

Chemicals	Composition (%wt.)				
	Sample 1	Sample 2	Sample 3	Sample 4	Sample 5
PDMS	5.54	5.54	5.54	5.54	5.54
PMHS			2.00	4.00	6.00
Cure catalyst		0.50	0.50	0.50	0.50
Wetting agent	0.20	0.20	0.20	0.20	0.20
Hexane	94.26	93.76	91.76	89.76	87.76

2.3 Specimens preparation

1 g of release agent sample was sprayed on 2.5 x 5 cm of the cured butyl rubber sheet (as a model for bladder). Then, the coated sheet rubber was placed into a hot air oven (Memmert UN55) at 150°C for 20 min to dry and cure the release agent film.

2.4 Physical and mechanical measurements

The friction coefficient and film endurance test were carried out by a tribometer [4] using a reciprocating mode according to ASTM G133-05 [5]. Each sample was tested at a load of 500 g against 6 mm tungsten carbine ball. The test conditions were: 5 mm/s speed for 4 mm linear distance. Coating surfaces were examined under optical microscope (Olympus DSX1000).

Releasing ability of green tire and bladder was tested by repeating compression molding technique. Coated bladder samples were hot-pressed with green tire compound at a pressure of 2 MPa and temperature of 170°C for 5 min. Number of compression cycles was recorded until the surface of the specimens adhere to the green tire.

3. Results and discussion

It was observed that as the solvent appeared to substantially evaporate from the sample surface the release coatings appeared to dry to a glossy appearance.

3.1 Evaluation of surface coatings by tribometer

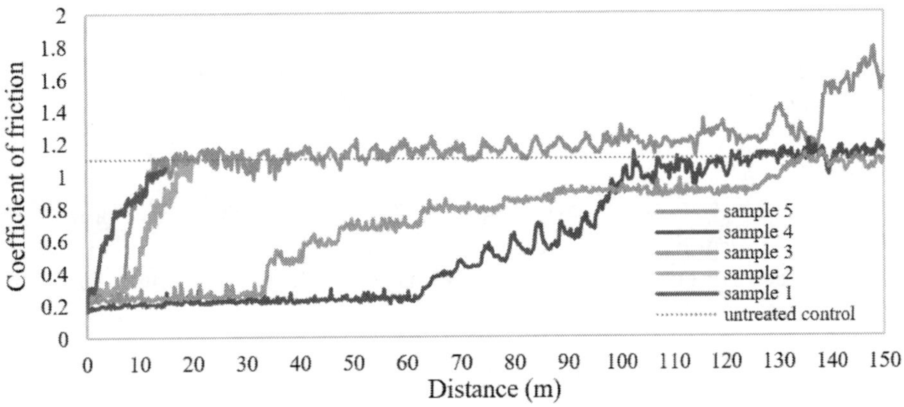

Figure 1. Coefficient of friction and film endurance of developed surface coatings tested by tribometer.

In the reciprocating mode in tribometer test, the sliding ball was moved on sample surface to monitor the frictional coefficient. Coating endurance can be determined from the transition of friction coefficient from stable to higher level [6]. It was found that the uncoated specimen (control) revealed high frictional coefficient of 1.1 due to rubbery surface of the substrate. As the samples were coated by release agents, frictional coefficient was decreased to the range of 0.25-0.30. By adding the metal salt catalyst (Sample 2 vs. Sample 1), film durability was improved. As the PMHS loading was increased, film endurance was increased. PMHS was converted into film to produce a resilient waterproof coating by using the metal salt catalyst at low temperature [7]. PMHS film shows good lubricating effect with outstanding water repellent property. Sample 1 had the least film endurance, whereas Sample 4 had the most film endurance as shown by the longest sliding distance.

IOP Conf. Series: Materials Science and Engineering 1234 (2022) 012005 doi:10.1088/1757-899X/1234/1/012005

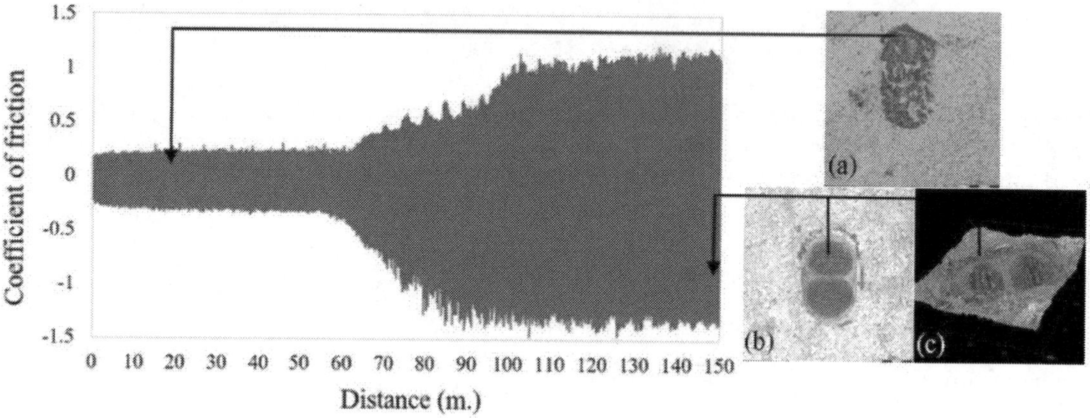

Figure 2. (a) Evaluation of friction coefficient vs. sliding distance on Sample 4 and corresponding optical micrographs showing wear tracks (b) at 20 m sliding distance and (c), (d) at 150 m sliding distance.

As shown in Figures 2a and 2b, a stable frictional coefficient of 0.25 at a distance of 0-62 m revealed existence of the release agent film. However, the coefficient of friction was raised to the value of 1.1 after 62 m meaning the tungsten carbide ball reached the rubber substrate as shown in Figures 2c and 2d.

3.2 Comparison of release agent efficiency

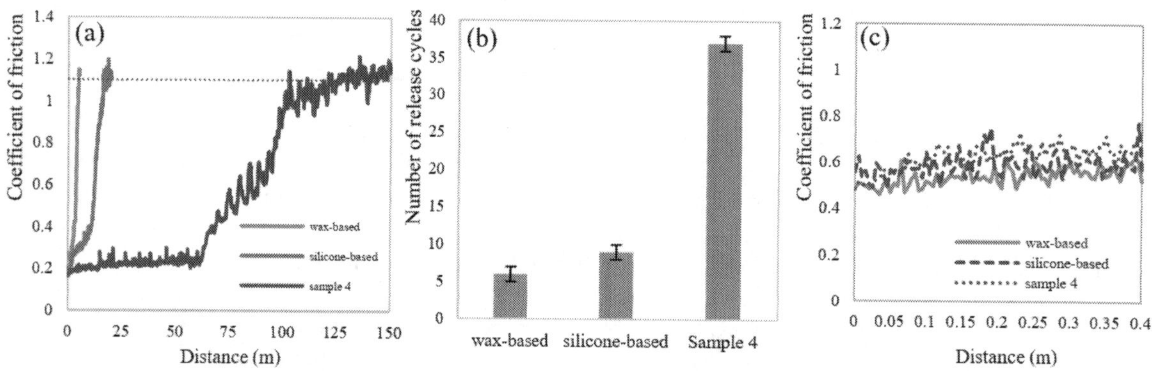

Figure 3. (a) coefficient of friction vs distance by tribometer (b) number of release cycles from repeating compression molding test and (c) coefficient of friction of each sample after failure.

In order to verify the release efficiency, comparative study of Sample 4 and other two commercial release agents (wax-based and silicone-based) were carried out. It was found that the commercial release agents had a higher coefficient of friction and lower stable distance value as shown in Figure 3a. The release efficiencies of each release agents were confirmed by repeating compression molding test. The number of release cycles by hot compression method of wax-based, silicone-based, and Sample 4 were 6, 9, and 37 cycles respectively [8] as shown in Figure 3b. Interestingly, it was found that the coefficient of friction on each sample after failure was in a range of 0.6-0.7 as shown in Figure 3c.

4. Conclusion

In this work, PDMS with PMHS release agent for bladder was developed. By adding the metal salt catalyst, film durability was improved. As the PMHS loading was increased, film endurance was increased. By using a reciprocating tribometer, friction coefficient of the release agent film was evaluated and tested for film durability. As confirmed by repeating compression molding test, efficiency of the release agents was estimated and used as a guideline for developing release agents in tire industry.

Acknowledgement

The research team is grateful to King Mongkut's Institute of Technology Ladkrabang (KMITL), King Mongkut's University of Technology North Bangkok (KMUTNB), and the National Science and Technology Development Agency (NSTDA) for co-financial supports. Thanks, are also to Acme International (Thailand) Ltd.

References

[1] H Brian 1994 *U.S.patent* no.0385919B1.
[2] F Louis 1982 *U.S.patent* USRE32318E.
[3] K Teerasurasassawat and P Koomsok 2018 Development of Mold Release Agent Formulas for Bladder in tire Industry., B.S. Thesis, KMITL, Thailand.
[4] W Gary 2006 *International journal of adhesion and adhesives.* **26** 577-599.
[5] ASTM G 133-05. 2005. Standard test method for linearly reciprocating ball-on-flat sliding wear.
[6] G Biresaw and Mittal K.L 2008. *Surfactants in tribology*, CRC Press, 101 pages.
[7] http://www.silicone-surfactant.com/ Methyl-Hydrogen-Silicone-Fluid.htm: November 2021.
[8] S Mahathanabodee 2021. Effects of Mold Release Agent on Adhesion of Molding Materials., National Science and Technology Development Agency, KMUTNB, Thailand.

IOP Conf. Series: Materials Science and Engineering 1234 (2022) 012006 doi:10.1088/1757-899X/1234/1/012006

Ultrafine fully vulcanized natural rubber modified by graft-copolymerization with syrene and acrylonitrile monomers

K Longsiri[1], P Mora[1], C Jubsilp[2], K Hemvichian[3] and S Rimdusit[1,] *

[1] Research Unit in Polymeric Materials for Medical Practice Devices, Department of Chemical Engineering, Faculty of Engineering, Chulalongkorn University, Bangkok 10330, Thailand; nurse.krittaphorn@gmail.com (K. L.); Phattarin.M@gmail.com (P. M.)

[2] Department of Chemical Engineering, Faculty of Engineering, Srinakharinwirot University, Nakhonnayok 26120, Thailand; chanchira@g.swu.ac.th

[3] Thailand Institute of Nuclear Technology, Nakhonnayok 26120, Thailand; kasineeh@yahoo.com

* Corresponding author: Sarawut.r@chula.ac.th

Abstract. Natural rubber (NR) is known as one of the most important renewable agricultural resources in Thailand. To enhance the NR's value, development and utilization of NR in form of commercial product are necessary. Commercial product based on ultrafine fully vulcanized powdered rubber (UFPR) was widely used in numerous potential applications such as toughening filler or friction modifier in brake pads. However, to meet the required thermal stability of friction modifier in brake pads for replacing the synthetic rubber powder, the modification of NR by graft-copolymerization with styrene (ST) and acrylonitrile (AN) monomer and production into the powder form are promised related to a sustainable economic system. From the beginning, the grafting process of NR latex was prepared by emulsion graft-copolymerization with ST and AN monomers onto deproteinized natural rubber (DPNR). The results revealed that the grafting with ST/AN monomer content of 80/20 weight ratio possessed the monomer conversion and grafting efficiency as high as 71% and 63%, respectively. After that, the obtained DPNR-g-(PS-co-PAN) was radiated by an electron beam, followed by a spray drying process to produce ultrafine fully vulcanized powdered natural rubber (UFPNR-g-(PS-co-PAN)). The SEM micrographs showed that UFPNR particles are relatively spherical with a particle size of approximately 3.56 μm. Furthermore, thermal stability i.e., degradation temperature at 5 % weight loss (T_{d5}) of the modified UFPNR was substantially improved from 334 to 350°C. The benefits of these advanced technologies give an obvious conclusion proved that the modified UFPNR is a good candidate for replacing a commercial synthetic powdered rubber and suitable for using as toughening filler in high performance polymer composite applications.

Keywords: Graft copolymer; Natural rubber-graft-polystyrene-co-polyacrylonitrile; Ultrafine fully vulcanized powdered natural rubber; Electron beam vulcanization; Spray drying.

1. Introduction

Natural rubber (NR) has been playing a very important role in rubber industry. Extending its application by producing into a powdered form which is, ultrafine fully vulcanized powdered natural rubber

IOP Conf. Series: Materials Science and Engineering 1234 (2022) 012006 doi:10.1088/1757-899X/1234/1/012006

(UFPNR), can be used as a toughening modifier in polymer composite or friction modifier in automotive friction materials. Due to the very high crosslinking degree on the surface of rubber particles and a moderate crosslinking degree in the inner parts obtained from the radiation process provided good elasticity properties and is easily dispersed in polymer matrices when blended [1,2]. However, to produce NR powdered has to be modified by graft copolymerization with styrene (ST) and acrylonitrile (AN) to improve thermal stability and reduce tackiness between rubber particles. Moreover, the grafted NR can enhance the interfacial adhesion between NR and polymer matrices, therefore improving the blends' properties [3, 4].

In the present work is devoted to modifying UFPNR by graft-copolymerization with ST and AN monomers on to deproteinized natural rubber (DPNR). The grafted NR was radiated by an electron beam followed by spray drying process to produce UFPNR. The effects of monomer ratio and radiation doses on thermal stability and morphology of UFPR were investigated.

2. Experimental

Effect of ST/AN contents on monomer conversion and grafting efficiency of graft copolymerization

The NR has a highly cis-1,4- polyisoprenic structure. It possesses labile allylic protons in every repeating unit, which can facilitate by chemical modifications [5]. In the grafting process of DPNR, a grafting position was generated on the NR surfaces, and monomer radicals can react to polyisoprenyl macroradicals via hydrogen abstraction reaction leading to grafting [6]. Therefore, the percentage by weight of ST/AN grafting onto NR at various ratios, i.e., 20/80, 30/70, 50/50, 70/30, and 80/20 were determined.

Monomer conversion and grafting efficiency dependency of monomer contents were measured by gravimetric method, according to equation (1) and (2), respectively [3]

$$\text{Monomer conversion (conv\%)} = \frac{\text{Weight of PS and PAN in gross copolymer}}{\text{Total weight of monomers}} \times 100 \quad (1)$$

$$\text{Grafting efficiency (GE\%)} = \frac{\text{Weight of polymer linked to natural rubber}}{\text{Total weight of polymer in gross copolymer}} \times 100 \quad (2)$$

As a result, shown in Figure 1. the monomer conversion and grafting efficiency significantly increased with the increasing ST content in the monomer mixture. The maximum value for the highest Conv and GE were achieved at weight ratio of 80/20 percentage by weight of ST/AN. Since, ST has similar polarity to polyisoprene in NR, which has higher reactivity to firstly react with polyisoprenyl macroradicals capable of further copolymerizing with AN monomer [4,7]. As the grafting on natural rubber can permanently retain the modified surface of NR particles by reducing the tackiness between particles. Therefore, modified NR will be beneficial for the better processing performance of NR powder.

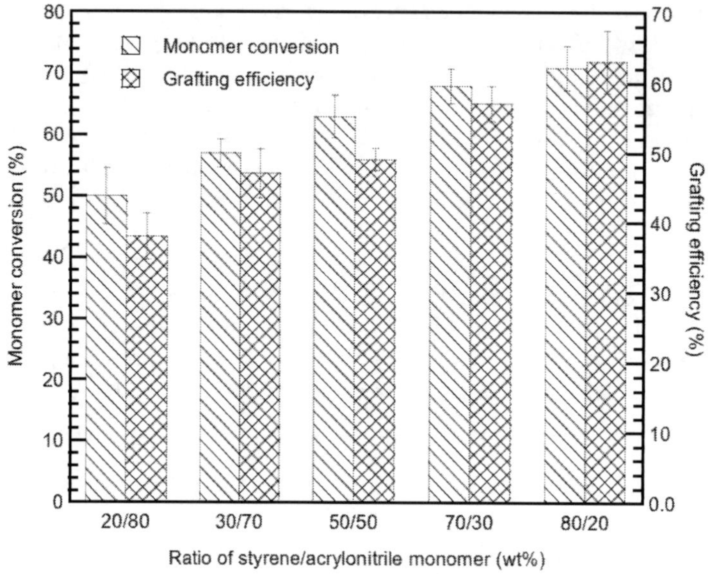

Figure 1. Variation of ST/AN percentage by weight as a function of (▧) monomer conversion and (▩) grafting efficiency.

3. Results and discussion

Effect of radiation doses on thermal stability of UFPNR

The modified UFPNR was prepared by radiation of DPNR-g-(PS-co-PAN) with an electron beam under various absorbed doses i.e., 100, 200, and 300 kGy, followed by a spray drying process. The grafted NR with ST/AN at 80/20 percentage by weight provided the highest monomer conversion and grafting efficiency was selected to investigate the degradation temperature at 5% weight loss (T_{d5}). Figure 2a). exhibited TGA thermograms of virgin NR and UFPNR-g-(PS-co-PAN) at various radiation doses. T_{d5} of grafted NR was found to be higher than virgin NR due to the presence of PS-co-PAN, besides the stable aromatic rings in PS may prevent the formation of macroradicals in NR structure when exposed to heat [7,8].

The effects of electron beam radiation with the rubber are the formation of free radicals, which induce to chain crosslinking or chain scission in the rubber chain. As the TAG graph, the T_{d5} of modified UFPNR was increased from 340 to 350°C with increasing the radiation doses. Due to the absorbed energy could be promoted more crosslinking than chain scission. Since, the presence of unsaturated groups in the rubber chain can bring about thermally induced crosslinking reactions, the crosslink density of UFPNR was obtained during the radiation process. Therefore, the higher crosslink density in the polymer chains can stabilize and restrict molecular mobility, more thermal energy may be required to cause destructive changes [9].

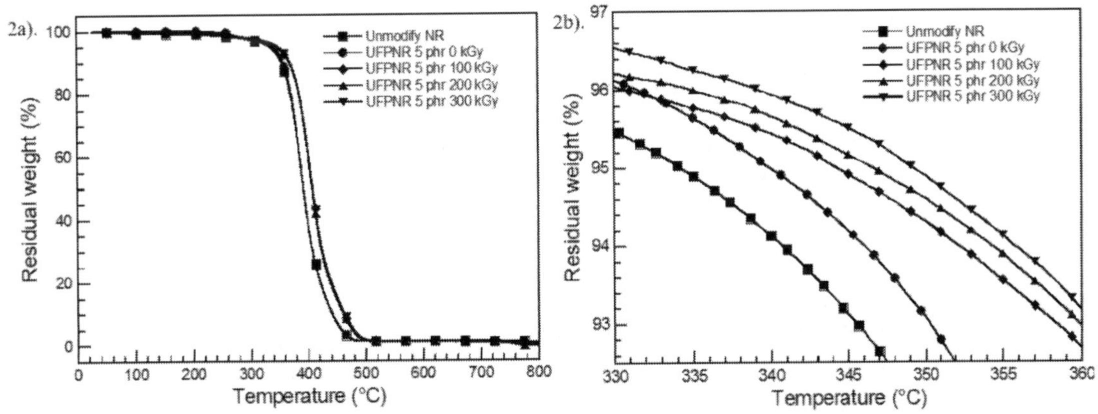

Figure 2. a) TGA thermogram of UFPNR-g-(PS-co-PAN) at various monomer contents and b) with an expanded axis.

Table 1. Degradation temperature at 5% weight loss of UFPNR.

Sample	ST/AN (wt.%)	Radiation dose (kGy)	T_{d5} (°C)
Virgin NR	-	Unirradiated	334
UFPNR-g-(PS-co-PAN)	80/20	Unirradiated	340
UFPNR-g-(PS-co-PAN)	80/20	100	344
UFPNR-g-(PS-co-PAN)	80/20	200	346
UFPNR-g-(PS-co-PAN)	80/20	300	350

Morphology of UFPNR

SEM investigated the morphology of UFPNR. Figure 3 a). and b). shows SEM micrographs of the virgin NR (unirradiated) and UFPNR (radiated at 300 kGy), respectively. The rubber particles tend to fuse and shown aggregation problems. Therefore, it is necessary to modified UFPNR by grafted with ST/AN. Fig 3 c). shows the morphology of modified UFPNR was radiated by electron beam at 300 kGy. The grafting and radiation process can provide spherical particles, most dispersed with a smooth surface and less aggregation. The higher degree of crosslinking after radiation may reduce the tackiness of the rubber particles [10]. Additionally, the average particle size of the modified UFPNR was approximately 3.56 ± 1.70 μm.

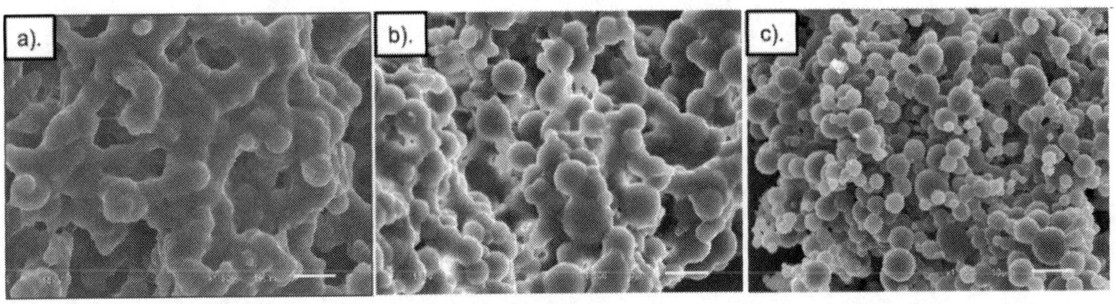

Figure 3. SEM micrographs (×1500 magnification) of a). virgin NR (unirradiated), b). UFPNR (radiated at 300 kGy) c). UFPNR-g-(PS-co-PAN) radiated at 300 kGy.

4. Conclusions

The monomer conversion and grafting efficiency of DPNR-g-(PS-co-PAN) as high as 71% and 63%, respectively with the increasing ST content. The modified NR was irradiated by electron beam radiation and spray drying to produce the UFPNR-g-(PS-co-PAN). The grafting and radiation process was significantly improved the T_{d5} of UFPNR. Furthermore, the increment of radiation doses up to 300 kGy provide smooth surface and lower agglomerated particles of rubber powder. The results revealed that thermal stability of the modified UFPNR grafted with ST/AN copolymers was significantly improved suitable for using as toughening filler in high performance polymer composite applications.

Acknowledgements

The authors would like to express their sincere appreciations to The National Research Council of Thailand (NRCT) and Thailand Science Research and Innovation (TSRI) for financial support throughout the research. Electron beam radiation was supported by Thailand Institute of Nuclear Technology (Public Organization).

References

[1] Wang, J., Zhang, X., Jiang, L., and Qiao, J., *Progress in Polymer Science*, 2019. 98: p. 101-160.
[2] Qiao, J., *Advanced Industrial and Engineering Polymer Research*, 2020. 3(2): p. 47-59.
[3] Okaniwa, M., and Suzuki, M., *Journal of Applied Polymer Science*, 2001. 81(14): p. 3462-3470.
[4] Angnanon, S., Prasassarakich, P., and Hinchiranan, N., *Polymer-Plastics Technology and Engineering*, 2011. 50(11): p. 1170-1178.
[5] Kovuttikulrangsie, S., Sahakaro, K., Intarakong, C., and Klinpituksa, P., *Journal of Applied Polymer Science*, 2004. 93(2): p. 833-844.
[6] Oliveira, P.C., Oliveira, A.M., Garcia, A., Souza Barboza, J.C., Carvalho Z., Cecília A., and Santos, A.M., *European Polymer Journal*, 2005. 41(8): p. 1883-1892.
[7] Nguyen Duy, H., Rimdusit, N., Tran Quang, T., Phan Minh, Q., Vu Trung, N., Nguyen, T.N., Nguyen, T.H., Rimdusit, S., Ougizawa, T., and Tran Thi, T., *Polymers for Advanced Technologies*, 2020. 32(2): p. 736-747.
[8] Nguyen H., Q.D., Tran V., Dung A., and Kawahara, S., *Polymers for Advanced Technologies*, 2019. 31.
[9] Bandzierz, K.S., Reuvekamp, Louis A. E. M., Przybytniak, Grażyna, D., Wilma K., Blume, A., and Bieliński, D.M., *Radiation Physics and Chemistry*, 2018. 149: p. 14-25.
[10] Rezaei Abadchi, M., and Jalali-Arani, A., *Nuclear Instruments and Methods in Physics Research Section B: Beam Interactions with Materials and Atoms*, 2014. 320: p. 1-5.

IOP Conf. Series: Materials Science and Engineering 1234 (2022) 012007 doi:10.1088/1757-899X/1234/1/012007

Influence of nanofiller types on morphology and mechanical properties of natural rubber nanocomposites

B Wongvasana[1], A Masa[2], H Saito[3], T Sakai[4] and N Lopattananon[1,*]

[1]Department of Rubber Technology and Polymer Science, Faculty of Science and Technology, Prince of Songkla University, Pattani 94000, Thailand.
[2]Rubber Engineering & Technology Program, International College, Prince of Songkla University, Songkhla, 90110, Thailand.
[3]Department of Organic and Polymer Materials Chemistry, Tokyo University of Agriculture and Technology, Koganei-shi, Tokyo 184-8588, Japan.
[4]Tokyo Office, Shizuoka University, 3-3-6 Shibaura, Minato, Tokyo, 108-0023, Japan.

*Corresponding author: natinee.l@psu.ac.th

Abstract Natural rubber (NR) nanocomposite containing different types of filler, i.e., nanoclay (clay) and cellulose nanofiber (CNF) were prepared in this study. The masterbatches of NR with 5 parts per hundred parts of rubber of nanofiller were firstly prepared by using the latex mixing method, followed by compounding on two roll mill and compression molding to obtain NR nanocomposite specimens. The unfilled NR sample was also prepared for comparison. Morphological properties of NR nanocomposites were investigated by using transmission electron microscopy, while the mechanical and dynamic properties were studied by using a universal tensile testing machine and dynamic mechanical analyzer (DMA). It was found that the clay with platelet morphology was uniformLy dispersed, while the long and flexible CNFs were aggregated and poorly dispersed. The greater improvement of modulus at various strains was achieved from CNF filled NR nanocomposites, while the highest tensile strength was obtained from the clay filled nanocomposite. As compared to the clay containing nanocomposite, the addition of CNF markedly decreased the tensile strength and elongation at break of the NR due to poor dispersion of CNF. However, a significant improvement in mechanical properties at low strain was obtained when the CNF was used as filler due to high degree of fiber entanglement, as suggested by DMA observation.

1. Introduction

Natural rubber (NR) is a flexible material used extensively in a variety of applications such as tires, dipped goods, adhesives, rubber thread, foam, conveyor belts, hoses, gaskets, footwear, and engine mounts, because of its unique characteristic. However, the NR is rarely used in its pure form as it is too weak to fulfil a practical applicationrequirement due to lack of hardness, strength and wear resistance. Basically, one of classical ways to achieve property improvement of NR is simply accomplished by adding reinforcing fillers.

In the past, when the NR was reinforced with nanofillers such as nanoclay (clay), graphene, carbon black, and cellulose nanofiber (CNF), etc. showed a great potential for producing a wide variety of high value products. Among these materials, the clay and CNF have been receiving a lot of attention in the

research and industrial sectors due to their economical and environmental attentions. However, the comparative study of the NR nanocomposites reinforced with clay and CNF are limited.
To understand the influence of filler geometry on morphology and mechanical properties of NR nanocomposites, the NR filled with clay or CNF were prepared. In previous studies [1,5], it was found that the maximum property enhancement was attained at 5 parts per hundred of rubber (phr) for both types of filler. Therefore, in this study, the NR was mixed with two different nanofillers at a fixed concentration 5 phr to fabricate the NR nanocomposites. The experimental observations on the microstructure of the clay/NR and CNF/NR nanocomposites as well as their mechanical properties were compared. The unfilled NR was also prepared as a reference.

2. Experimental

Two different kinds of NR masterbatches were prepared via the latex mixing method. Suspension of clay (Kunipia-F®, Kunimine Industries Co. Ltd., Tokyo, Japan) or CNF (Nanoforest-S, Chuetsu Pulp and Paper Co. Ltd., Tokyo, Japan) in water was added into NR latex (HA-NR latex, Yala Latex Co. Ltd., Yala, Thailand). The NR and nanofiller were mixed thoroughly under vigorous stirring at ambient temperature for 30 min, and dried at 50°C for 3 days to obtain the clay/NR and CNF/NR masterbatches containing 5 phr nanofillers. The unfilled NR was also prepared using the same procedure and used as a reference. For the preparation of NR and their nanocomposites, the dried NR and masterbatches of NR nanocomposites were masticated and mixed with paraffinic oil (China Petrochemical International Co. Ltd., Shanghai, China), and 2,2,4-trimethyl-1,2-dihydroquinone (TMQ, Lanxess AG, Cologne, Germany) in an internal mixer (CT internal mixer, Chareon TUT Co., Ltd., Samutprakan, Thailand) at a temperature of 50°C with a rotor speed of 60 rpm for 10 min. The obtained NR and NR nanocomposite compounds were mixed with dicumyl peroxide (DCP, Wuzhou International Co. Ltd., Shenzhen, China) on a two-roll mill. The compounds were then hot-pressed at 90% cure in a mold for 8 min at 160°C. the samples were cut into test specimens for microscopic analysis and mechanical property measurement.

3. Results and discussion

Dispersion of nanofillers (clay and CNF) in the NR nanocomposites is shown in Figure 1(A) and (B).From Figure 1(A), it is seen that the clay particles with intercalated structure where the NR chains intercalated into interlayer space of the clay tactoids [1], shown as a dark line, were homogeneously dispersed in the NR matrix (lighter region). Moreover, some of the clay remained as aggregated particles or tactoids having average thickness and length of about 22 ± 16 nm and 158 ± 87 nm, respectively. In the case of CNF/NR nanocomposite (Figure 1(B)), there were different grades of CNF aggregations, ranging from small aggregate (i.e., fiber-liked structure) to large aggregate (i.e., spherical-liked one). The average thickness of CNFs was about 132 ± 74 nm while the average length was approximately 1-3 μm. It is also observed from the TEM image (Figure 1 (B)) that the CNFs were highly entangled. The CNFs are long and flexible nanofibers, and they have a high density of hydroxyl groups present on their surfaces, therefore the CNFs are easily entangled and interact through hydrogen bonding, leading to forming CNF aggregates of different shapes and dimensions [2-5].

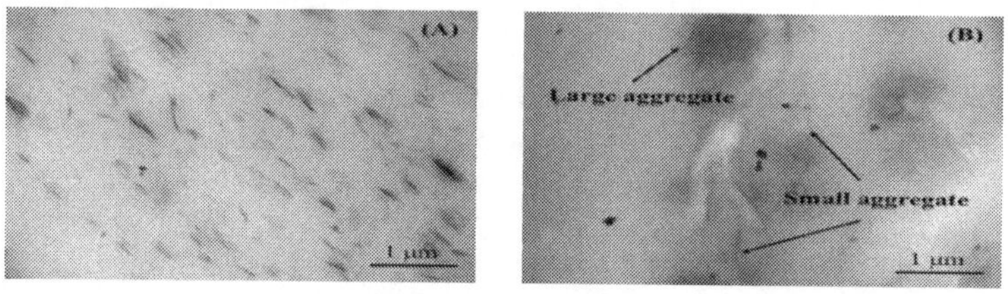

Figure 1. TEM images of NR nanocomposites: (A) clay/NR, and (B) CNF/NR nanocomposites.

Figure 2. shows the stress-strain curves of the NR and its nanocomposites, and the values of modulus at 100% strain, tensile strength and elongation at break are tabulated in Table 1. It is seen that the inclusion

IOP Conf. Series: Materials Science and Engineering 1234 (2022) 012007 doi:10.1088/1757-899X/1234/1/012007

of nanofiller influenced the stress-strain behaviour of NR. The addition of clay altered the stress-strain curves of NR by a slowly increase of stress at low strain and sharply increase of stress at high strain due to collaborative strain-induced crystallization [1]. On the other hand, the CNF/NR nanocomposite did not show any evidence of such process during tensile testing. As compared to the unfilled NR, the addition of clay slightly increased the stiffness (i.e., Young's modulus and 100% modulus), and considerably increased tensile strength. The increased stiffness caused a reduced flexibility of the clay/NR nanocomposite. The good dispersion of clay in the NR matrix would be responsible for this enhancement. The addition of CNF markedly decreased both the tensile strength and elongation at break of the NR due to large aggregation and poor dispersion of CNF in the NR matrix, as already shown in Figure 1(B). Therefore, the large CNF aggregate acted as a rubber network defect, which lowered the mechanical properties of the NR [6-7]. However, it is very interesting to mention that the incorporation of CNF significantly enhanced the Young's modulus and 100% modulus of the NR. The enhancement was about 119% and 150% over that of the clay/NR nanocomposite, respectively. It was believed that the entanglement between CNF-CNF and CNF-NR imparted the rigidity of the NR nanocomposite, which increased the Young's modulus and 100% modulus of NR. This explanation was later confirmed by DMA result.

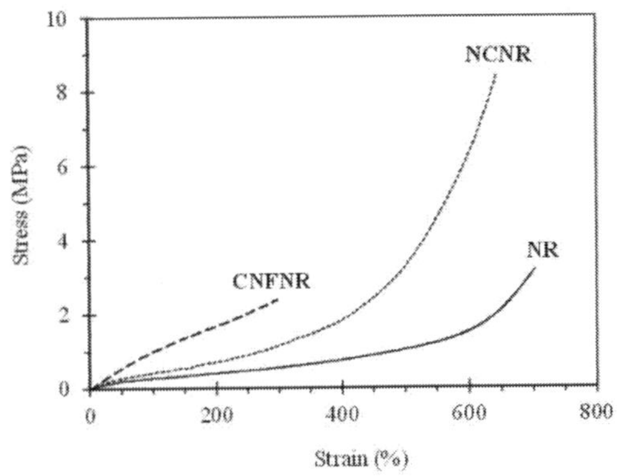

Figure 2. Stress-strain curves of NR and their nanocomposites.

Table 1. Young's modulus, 100% modulus, tensile strength, and elongation at break of NR and their nanocomposites.

Samples	Young's modulus (MPa)	100%Modulus (MPa)	Tensile strength (MPa)	Elongation at break (%)
NR	0.0033	0.24±0.03	3.64±0.85	759±16
Clay/NR	0.0054	0.35±0.01	9.81±0.39	672±11
CNF/NR	0.0118	0.90±0.03	2.38±0.16	293±11

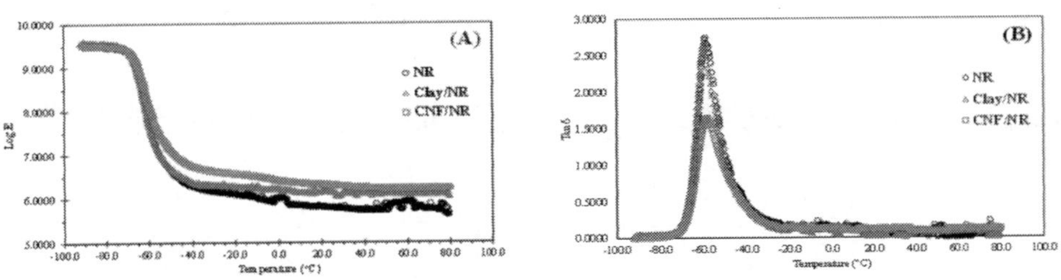

Figure 3. Variation of (A) storage modulus (log G′) and (B)loss tangent (Tan δ) as a function of temperature for NR and their nanocomposites.

Table 2. Storage modulus (G′) at 25°C, degree of entanglement and maximum loss tangent (Tan δ_{max}) of NR and their nanocomposites.

Samples	Log G′ at 25°C(MPa)	Degree of entanglement(mole/cm^3)	Tan δ_{max}
NR	5.79	4.623x10^{-5}	2.55
Clay/NR	6.13	7.181x10^{-5}	2.65
CNF/NR	6.30	1.476x10^{-4}	1.64

Figure 3 shows storage modulus (G′) andloss tangent (Tan δ) as a function of temperature for the NR and their nanocomposites. Their corresponding values, i.e., degree of entanglement are also tabulated in Table 2. The degree of entanglement was calculated by using; Degree of entanglement = E′/6RT [8], where E′ is the storage modulus obtained from the plateau region, R is the universal gas constant and T is the absolute temperature. From both Figure 3 and Table 2, the incorporation of both fillers increased G′ at 25°C and the improvement was even more evident for the NR reinforced with CNF (Figure 3(A)), indicating that higher stiffness was obtained from the CNF filled NR nanocomposite. In addition, the lowest Tan δ peak was found in the CNF/NR nanocomposite, indicating that the higher interactions at the interface between the NR matrix and CNF nanofillers. This was attributed to the CNF has higher surface area to interact with NR than clay. High interactions between NR and nanofiller would effectively restrict the mobility of the NR chains near the interface region, lowering the ability of sample to dissipation energy, resulting in lowering the peak height of Tan δ. From Table 2, it was seen that the degree of entanglement value was increased with adding nanofillers, and the increment was more pronounced when the CNF was used as a nanofiller. The higher degree of entanglement in the CNF/NR nanocomposite further substantiated the increase of storage modulus and mechanical properties in the CNF/NR nanocomposites at low strain [9-10].

4. Conclusions

The nanocomposites containing 5 phr nanofiller were successfully prepared by using the latex mixing method. Based on the obtained results, it was found that the clay with platelet morphology was uniformly dispersed, while the long and flexible CNFs were aggregated and poorly dispersed. The greater improvement of modulus at various strains was achieved from the 5 phr CNF filled NR nanocomposites, while the highest tensile strength was obtained from the 5 phr clay filled nanocomposite. The addition of CNF markedly decreased the tensile strength and elongation at break of the NR due to poor dispersion of CNF. However, a significant improvement in 100% modulus was obtained from the 5 phr CNF/NR nanocomposite due to higher degree of entanglement in CNF/NR nanocomposite, as suggested by DMA.

Acknowledgement

We gratefully acknowledge financial supports from the Thailand Research Fund (TRF) through the Royal Golden Jubilee Ph.D. Program (Grant No. PHD/0156/2560), the Thailand Science Research and Innovation (TSRI) and the National Research Council of Thailand (NRCT).

References

[1] Masa A, Saito R, Saito H, Sakai T, Kaesaman A and Lopattananon N 2015 *J. Appl. Polym. Sci.* **133**, 43214.
[2] Dufresne A 2013 *Mater. Today*. **16**, 220.
[3] Thomas MG, Abraham E, Jyotishkumar P, Maria HJ, Pothen LA and Thomas S 2015 *Int. J. Biol. Macromol.***81**, 768.
[4]Panesar D, Leung R, Sain M and Panthapulakkal S. In *Sustainable and Nonconventional Construction Materials Using Inorganic Bonded Fiber Composites*; Junior, H. S.; Fiorelli, J.; Dos Santos, S. F., Ed.; Woodhead Publishing., Sawston, UK, pp. 17-26.
[5] Fiorote JA, Freire AP, Rodrigues DS, Martins MA, Andreani L and Valadares LF 2019 *Bioresources*. **14**, 3168.

[6] Lopattananon N, Panawarangkul K, Sahakaro K and Ellis BJ 2006 *J. Appl. Polym. Sci.***102**, 1974.
[7] Lopattananon N, Jitkalong D and Seadan MJ 2011 *J. Appl. Polym. Sci.* **120** 3242.
[8] Komalan C, George K, Kumar P, Varughese K and Thomas S 2007 *Express.Polym. Lett.* **1**, 641.
[9] Xu X, Liu F, Jiang L, Zhu JY, Haagenson D and Wiesenborn DP 2013 *ACS Appl. Mater. Interfaces.* **5**, 2999.
[10] Balakrishnan P, Gopi S, Geethamma VG, Kalarikkal N and Thomas S 2018 *Macromol. Symp.***380**, 1.

IOP Conf. Series: Materials Science and Engineering 1234 (2022) 012008 doi:10.1088/1757-899X/1234/1/012008

Thermoplastic natural rubber based on linear-low-density polyethylene

R Yoksan[1,2,*] and W Wannawitayapa[1]

[1] Department of Packaging and Materials Technology, Faculty of Agro-Industry, Kasetsart University, Bangkok 10900, Thailand
[2] Center for Advanced Studies for Agriculture and Food, Kasetsart University Institute for Advanced Studies, Kasetsart University, Bangkok 10900, Thailand

*Corresponding author: rangrong.y@ku.ac.th

Abstract. Thermoplastic natural rubber (TPNR) is one of the alternative natural rubber (NR) products obtained by blending NR with any compatible thermoplastics, e.g., high-density polyethylene (HDPE), low-density polyethylene (LDPE), linear-low-density polyethylene (LLDPE), polystyrene (PS), polylactic acid (PLA) at the temperature above the melting point of plastics, which is always above the softening point of NR. Unlike vulcanized NR, TPNR is recyclable and convertible to the final articles using the existing plastic technologies, including extrusion and injection molding. TPNR materials based on NR/LLDPE blends filled with two types of inorganic fillers, i.e., titanium dioxide and silica, were prepared and their properties were discussed in the current work. Silica provided a better reinforcing effect to the NR/LLDPE blends, whereas the blends filled with titanium dioxide exhibited excellent UVA and weather resistance. The resulting NR/LLDPE blends filled with inorganic fillers can be injection molded to form the products with desired shapes.

1. Introduction

Natural rubber (NR) is a biobased polymeric material from the Para rubber tree (Hevea brasiliensis). NR has high molecular weight and exhibits excellent water resistance, but low grease resistance. Furthermore, NR is sensitive to oxidation because of unsaturated double bonds; metal, light, and heat can accelerate the oxidation of NR as well. Consequently, NR properties may be changed over time during outdoor uses.

Thermoplastic natural rubber (TPNR) is a kind of thermoplastic elastomer, which is one of the alternative NR products obtained by blending NR with any compatible thermoplastics, e.g., high-density polyethylene (HDPE) [1], low-density polyethylene (LDPE) [2], linear-low-density polyethylene (LLDPE) [3], polystyrene (PS) [4], polylactic acid (PLA) [5] at the temperature above the melting point of plastics, which is always above the softening point of NR. Unlike vulcanized NR, TPNR is recyclable and convertible to the final articles using the existing plastic technologies, including extrusion and injection molding.

Herein, TPNR materials based on NR/LLDPE blends filled with two types of inorganic fillers, i.e., titanium dioxide and silica were prepared and their properties, including mechanical, morphological, and QUV and weather-resistant ones were discussed.

2. Experimental

2.1. Preparation of NR/LLDPE blends filled with TiO_2 and/or SiO_2

Natural rubber (NR, STR 5CV60, Natural Art and Technology Co., Ltd., Thailand) was first masticated in an internal mixer (KD-3-20, Kneader Machinery, Japan) at 70°C for 2 min. Antioxidant and antiozonant (1 phr each, Behn Meyer Chemicals (Thailand) Co., Ltd., Thailand) were added into the masticated NR and their mixing was performed at 70°C for 2 min. The mixture was then cut into small pieces using a plastic crusher to obtain NRC (NR without the addition of inorganic fillers). On the other hand, NR, antioxidant (1 phr), and antiozonant (1 phr) were compounded with inorganic fillers (20 wt%), i.e., TiO_2 (TIOXIDE® TR92, Venator Materials Corporation Huntsman, Malaysia), SiO_2 (TOKUSIL® 255, OSC Siam Silica Co., Ltd, Thailand), or TiO_2/SiO_2 (1:1 w/w) in an internal mixer at 70°C for 5 min. These compounds were prepared into small pieces using the same processes as those for NRC to obtain NRT (NR containing TiO_2), NRS (NR containing SiO_2), and NRTS (NR containing equal amount of TiO_2 and SiO_2), respectively.

The as-prepared NRT, NRS, or NRTS flakes were subsequently blended with NRC flakes and linear-low-density polyethylene (LLDPE, InnoPlus LL8420A, PTT Global Chemical Public Company Limited, Thailand) in an internal mixer at 110–140°C for 5 min to obtain NR/LLDPE/TiO_2, NR/LLDPE/SiO_2, and NR/LLDPE/TiO_2/SiO_2, respectively. A series of NR/LLDPE/TiO_2, NR/LLDPE/SiO_2, and NR/LLDPE/TiO_2/SiO_2 was prepared by varying weight fractions of the above three components. The final samples possessed a constant NR/LLDPE weight fraction of 60/40 and different concentrations of TiO_2, SiO_2, or TiO_2/SiO_2 (1/1 w/w) of 2.5, 5, and 10 wt.%. Each sample was then compressed at 100°C by a compression molding machine (CC-HIM-2060, Chaicharoen, Thailand) and then die cut into test specimens.

2.2. Characterization and property testing of NR/LLDPE blends filled with TiO_2 and/or SiO_2

The tensile property was tested according to ISO37:2011 using an Instron universal testing machine (Instron, USA) with a test speed of 500 mm/min and a gauge length of 30 mm. Tensile strength, Young's modulus, and elongation at break of each sample were reported as averages from the measurement of three specimens. Shore A hardness of the sample was tested according to the ASTM D2240-05 using a GS-719N Durometer (Teclock, Japan).

Morphological characteristics at tensile fractured surfaces of the samples were examined using a JSM-6610 series scanning electron microscope (SEM, JEOL, USA) at an acceleration voltage of 10 kV. The SEM micrographs were taken at a magnification of 500. The sample surfaces were coated with a thin layer of gold to prevent charging before observation.

Ozone resistance test was performed according to ISO1431-1:2004. Samples were stretched to 120% of their original length and then exposed to the ozone with a concentration of 50 pphm at 40°C for 7 h. The crack of the samples was then observed.

QUV and weathering resistance determination was carried out according to ASTM G154-16. The sample specimens were exposed to UVA at 60°C for 8 h and punctuated with 50°C of water vapor for 4 h for the overall test period of 168 h using an Accelerated Weathering Tester (QUV/spray, Q-LAB, USA). The color of each specimen was then examined using a spectrophotometer (Miniscan EZ, HunterLab, USA) and the change in color as compared with the unexposed specimen was observed.

The results were statistically analyzed using IBM SPSS Statistics Version 20 (Statsoft, Oklahoma), with analysis of variance (ANOVA) and Duncan's test at significant differences at 5% level.

3. Results and discussion

The tensile properties and hardness of all samples were shown in Figure 1. Although tensile properties and hardness of NR/LLDPE blend and NR/LLDPE/inorganic filler composites were insignificantly different, tensile strength, Young's modulus, and hardness of the composites tended to increase with increasing inorganic filler concentration. It should be pointed that elongation at break of the composites was lower than that of the blend and hardly changed as a function of inorganic fillers concentrations. The results implied that the composites became more rigid due to the reinforcing effect.

SEM images at tensile fracture surfaces also supported the above mechanical property results as demonstrated in Figure 2. NR/LLDPE blend shows rough tensile fractured surface, reflecting extensible material. However, the roughness of tensile fractured surfaces slightly decreased when inorganic fillers were added, corresponding to stiffer material. The reduction of surface roughness was more outstanding in the composite filled with SiO_2, particularly with increasing SiO_2 concentration.

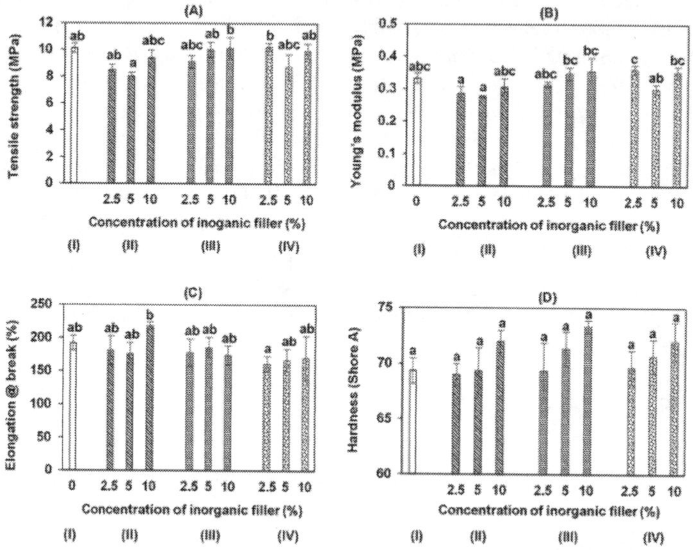

Figure 1. (A) Tensile strength, (B) Young's modulus, (C) elongation at break, and (D) hardness of (I) NR/LLDPE, (II) NR/LLDPE/TiO_2, (III) NR/LLDPE/SiO_2, and (IV) NR/LLDPE/TiO_2/SiO_2.

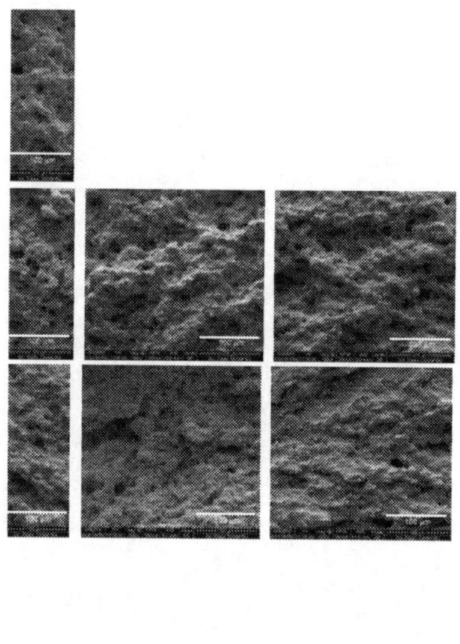

Figure 2. SEM images (×500) at tensile fractured surfaces of (I) NR/LLDPE, (II) NR/LLDPE/TiO_2, (III) NR/LLDPE/SiO_2, and (IV) NR/LLDPE/TiO_2/SiO_2.

QUV and weathering resistance properties of the blend and composites were evaluated by the difference in CIELAB color space value before and after exposure to the UVA, water vapor, and heat. The smaller difference indicates the better resistances. Figure 3A-C shows that the difference in CIELAB color space (L^*, a^*, and b^*) values of the composites filled with TiO_2 was lower than those of the blend and the composites filled with SiO_2, suggesting that TiO_2 provided the best resistance against UVA and weather to the NR/LLDPE blend. The result in Figure 3D was also in good agreement with the CIELAB color space as the composites filled with TiO_2 possessed a higher Grey Scale value than others, reflecting less change in color after UVA and water vapor exposure.

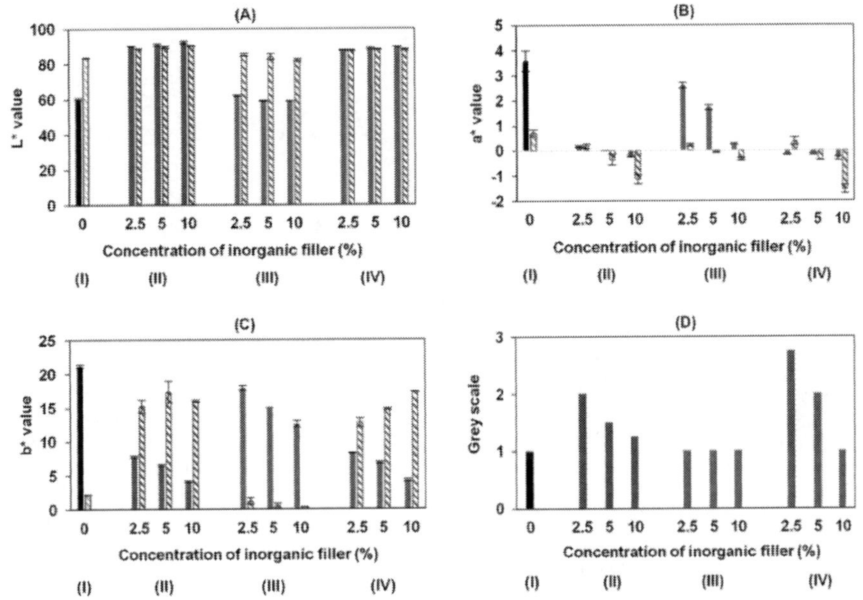

Figure 3. (A) L^*, (B) a^*, and (C) b^* values before (left bar) and after (right bar) exposing to UVA at 60°C for 8 h alternating with water vapor at 50°C for 4 h, total exposure time of 168 h, and (D) Grey Scale of (I) NR/LLDPE, (II) NR/LLDPE/TiO_2, (III) NR/LLDPE/SiO_2, and (IV) NR/LLDPE/TiO_2/SiO_2.

4. Conclusions

TPNR materials based on LLDPE were successfully prepared. Two types of inorganic fillers, i.e., TiO_2 and SiO_2, were incorporated. The obtained NR/LLDPE composites filled with inorganic additives showed slightly increased tensile strength, Young's modulus, and hardness, which were in good agreement with the morphology observed by SEM. SiO_2 provided the composites with better mechanical properties, whereas TiO_2 improved their QUV and weather resistance.

Acknowledgment

This work was financially supported by the Thailand Research Fund, Thailand, and the Graduate Program Scholarship from The Graduate School, Kasetsart University, Thailand.

References

[1] Akhtar, S., P.P. De, and S.K. De. J. Appl. Polym. Sci. 1986. 32(3): p. 4169-4183.
[2] Mahapram, S. and S. Poompradub. Polym. Test. 2011. 30(7): p. 716-725.
[3] Abdullah, I., S. Ahmad, and C.S. Sulaiman. J. Appl. Polym. Sci. 1995. 58(7): p. 1125-1133.
[4] Asaletha, R., M.G. Kumaran, and S. Thomas. Polym. Degrad. Stab. 1998. 61(3): p. 431-439.
[5] Nematollahi, M., A. Jalali-Arani, and H. Modarress. Polym. Int. 2019. 68(3): p. 439-446.

Study of parameter affecting morphology of electrospun poly (lactic acid) (PLA) fibers loaded with Ag/$CaCO_3$ filler

M E Zin[1,2] and W Sutapun[1,2,*]

[1] School of Polymer Engineering, Institute of Engineering, Suranaree University of Technology, Nakhon Ratchasima 30000, Thailand.
[2] Research Center for Biocomposite Materials for Medical Industry and Agricultural and Food Industry, Suranaree University of Technology, Nakhon Ratchasima 30000, Thailand.

*Corresponding author: wimonlak@sut.ac.th

Silver nanoparticles (AgNPs) are widely used as antimicrobial agent in commercial products like textiles, cosmetics and drugs. AgNPs deposited on the surface of calcium carbonate ($CaCO_3$) particles, serve as sustained release of antimicrobial activity. The silver nanoparticles embedded calcium carbonate particles (Ag/$CaCO_3$) are prepared by a precipitation method. In this study, precipitated (Ag/$CaCO_3$) filler will be used for preparation of Ag/$CaCO_3$-poly(L-lactic acid) (Ag/$CaCO_3$-PLA) nanofibers by electrospinning. Polymer concentration and functional filler amount in binary solvent system will be studied. The condition for fabricating electrospun PLA and Ag/$CaCO_3$-PLA was a 15 cm of collection distance and 15 kV of working voltage. Morphology of Ag/$CaCO_3$ loaded PLA electrospun fibers will be investigated by the field emission scanning electron microscope (FE-SEM). The electrospun fibers will be further applied as the antimicrobial material.

Keywords: PLA, Ag/$CaCO_3$ filler, Electrospinning, Antimicrobial Applications

1. Introduction

Silver nanoparticles (AgNPs) are widely used as antimicrobial agent in health care, food and commercial products like textiles. AgNPs have now been established as an effective biocidal agent and act on a wide range of both gram-negative and gram-positive bacteria [1]. AgNPs incorporated into polymer matrix are currently a popular type of nanocomposite that can provide excellent antimicrobial properties. Antimicrobial nano polymer composite could be applied as packaging films [2]. wound dressing materials [3] and other antimicrobial applications [4,5] according to their outstanding antimicrobial activity. To reduce agglomeration and maintain the nano size of Ag particles, calcium carbonate is one of a good choice as AgNPs carrier for incorporation into polymer matrix [6]. Polylactic acid (PLA) has been widely used in various applications including biomedical applications and packaging because of its biodegradability, biocompatibility and ability to be dissolved in common solvents for processing [7].

Electrospinning is simple and efficient for the fabrication of nano to microscale fibers. The morphology and diameter of the electrospun fibers depend on solution properties, processing and ambient factors [8]. In this research work, the electrospinning was employed to prepare mat of PLA and PLA loading with Ag/$CaCO_3$. Then, effects of PLA solution concentration, applied voltage, solution flow rate and Ag/$CaCO_3$ filler content on morphology of electrospun fibers and distribution of fiber diameter were studied. This research work is preliminary work for further study on application of Ag/$CaCO_3$ for food packaging.

2. Experimental

The commercial PLA grade (4043D) was purchased from NatureWorks. Chloroform RPE was purchased form Carlo Erba Reagents and acetone was purchased from Thermo Fisher Scientific. Silver nanoparticles loaded calcium carbonate particles ($Ag/CaCO_3$) were prepared by a coprecipitation method mentioned by *S. Zapotoczny* (2012) [6]. The particle size range of the $Ag/CaCO_3$ is around 2 to 5 μm with D(v,0.9) of 4.66 μm. Various concentrations of PLA were prepared with a binary solvent system of chloroform and acetone (chloroform: acetone = 2:1 in volume). The PLA solution with concentrations of 7.5%, 10%, 12.5% and 15% w/v were prepared by stirring at room temperature for 24 hrs using a magnetic stirrer. For preparing $Ag/CaCO_3$ filled PLA nanofibers, only 10% w/v PLA solution with the binary solvent system of chloroform and acetone was employed. Firstly, 3.00 g of PLA was dissolved in 20 mL of chloroform for 22 h and different $Ag/CaCO_3$ contents which were dispersed in 10 mL of acetone were added to the dissolved PLA solution. Then, the mixtures were magnetically stirred at room temperature for 2 h.

Each PLA solution, unfilled and filled, was loaded into a 10 mL plastic syringe (NIPRO syringe), and the needle (inner diameter 0.9 mm) was used for electrospinning. Two flow rates (1 mL/h and 1.5 mL/h) were varied and high voltage of up to 15 kV were applied. Electrospun fibers were collected on a grounded aluminum collector at a distance of 15 cm from the needle tip. The spinning time for pure PLA solutions and $Ag/CaCO_3$ loaded PLA solutions was 30 min and 15 min, respectively. All experiments were performed at about 30°C and relative humidity around 60%. Morphology of electrospun fibers made of pure PLA and $Ag/CaCO_3$ loaded PLA were investigated by the field emission scanning electron microscope (Carl Zeiss, Auriga) using an accelerating voltage of 3.0 kV. Before observation, each sample was coated with a gold sputter coater for 2 min (Neo Coater).

3. Results and Discussion

SEM micrographs in Figure 1 show morphology of the nanofibers electrospun with different PLA concentrations, flow rates and applied voltages. The SEM micrographs show characteristic of porous structure of PLA fibers electrospun from PLA solution with the binary solvent system of chloroform:acetone. The concentration of PLA solution does affect PLA nanofiber morphology and diameter. The micrographs show that smaller fibers with beads were collected from 7.5% w/v PLA solution (Figure 1A-D). The formation of beads indicates insufficient chain entanglements [9]. On increasing PLA concentration to 10%w/v, continuous fibers with various diameter fibers were produced (Figure 1E-H) and uniform fibers were collected at 12.5%w/v PLA solution (Figure 1I-L). At the highest PLA concentration of 15%w/v, electrospun fibers were folded back and forth not randomly oriented as observed from electrospun fiber with lower concentration of PLA solution (Figure 1M-P). Casasola *et al.* [9] have mentioned that increasing polymer concentration results in more chain entanglements, an increase in the solution viscosity. As a result, hence an increase in the viscoelastic force will be increased and will counterbalance the Coulombic stretching force. Therefore, it results in continuous fibers with fewer beads.

Uniform electrospun nanofibers in size could be prepared via a critical flow rate of a polymeric solution. At higher flow rate of 1.5 mL/h, fibers with larger diameter were produced comparing to the prepared fiber with lower flow rate of 1 mL/h. Increasing flow rate beyond a critical value not only leads to increase in the pore size and fiber diameter but also bead formation. This might be due to incomplete drying of the nanofiber jet during the flight between the needle tip and metallic collector [10]. Upon concerning applied voltage, smaller diameter of fibers was produced at higher applied voltage of 18kV compared to produced fibers at 15kV, as shown in Figure 1. The formation of smaller diameter nanofibers with an increase in the applied voltage is attributed to the stretching of the polymer solution in correlation with the charge repulsion within the polymer jet [10].

SEM micrographs in Figure 2 show electrospun fibers obtained from PLA solutions filled with 1 and 3% w/v $Ag/CaCO_3$ using 1 mL/h flow rate, and 18 and 20 kV applied voltages. At 1%w/v $Ag/CaCO_3$ loading, the electrospun filled PLA fibers contained some particles blooming along the fiber, as shown in Figure 2(A) and 2(B). However, with 3%w/v $Ag/CaCO_3$ loading, the electrospun fiber comparatively contained more particle blooming along the fiber due to higher degree of the local particle agglomeration

IOP Conf. Series: Materials Science and Engineering 1234 (2022) 012009 doi:10.1088/1757-899X/1234/1/012009

during electrospinning process, as illustrated in Figure 2(C) and 2(D). The local particle agglomeration might be caused by heterogeneous distribution of the particles in PLA solutions during electrospinning.

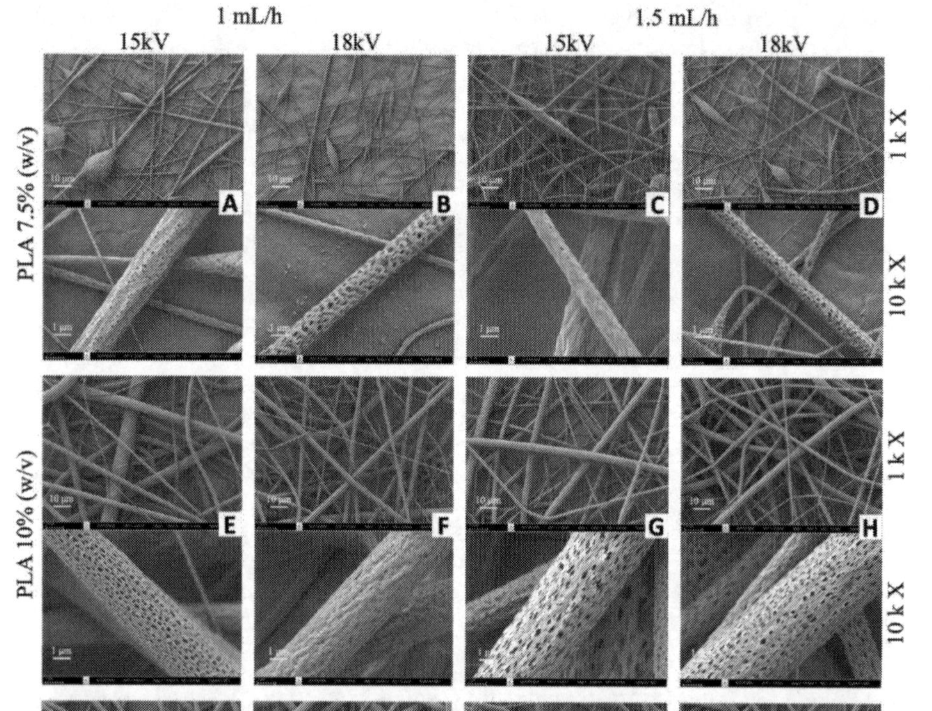

Figure 1. SEM micrographs of electrospun PLA fiber prepared from various concentrations, flow rates of 1 and 1.5 mL/h, and applied voltages of 15 and 18 kV.

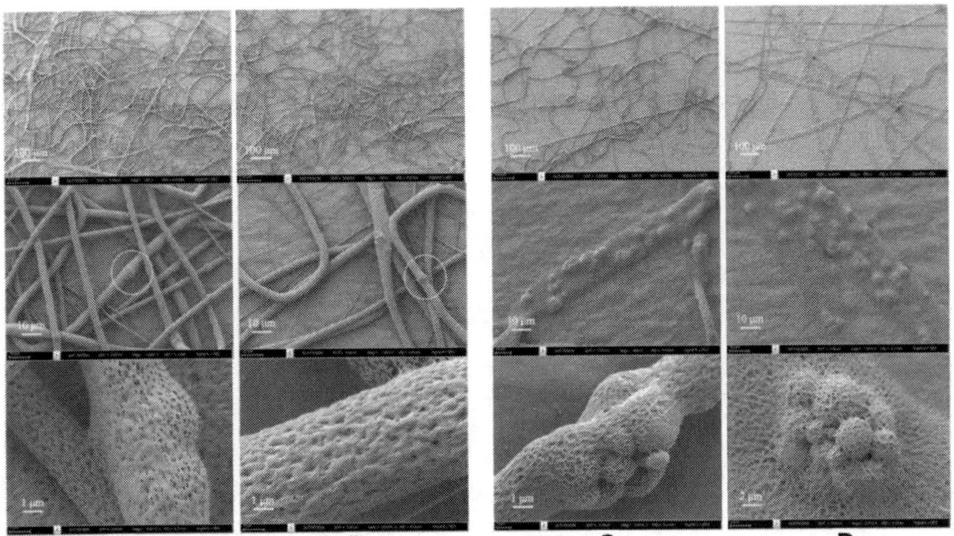

Figure 2. SEM micrographs of the electrospun filled PLA fiber with 1% (w/v) Ag/$CaCO_3$ at 1 mL/h and 18 kV (A), and 1mL/h and 20 kV (B); and 3% (w/v) Ag/$CaCO_3$ at 1 mL/h and 18 kV (C), and 1mL/h and 20 kV (D).

4. Conclusions

Highly porous fibers were prepared by electrospinning from pure PLA solution. PLA solutions of 10 and 15%w/v concentrations would give rise to electrospun continuous fibers without beads. Larger diameter electrospun fibers were obtained at higher flow rate of 1.5 mL/h and applying higher voltage of 18 kV resulted in smaller diameter fibers than applying lower voltage. PLA electrospun fibers filled with 1%w/v Ag/$CaCO_3$ contained some particles blooming along the fiber however, more particle blooming and nonuniform shape were observed from the electrospun fibers filled with 3%w/v Ag/$CaCO_3$. The electrospun fibers obtained from 1%w/v Ag/$CaCO_3$ loading with applied voltage of 20 kV were more uniform in shape.

Acknowledgement

The authors gratefully acknowledge the financial support from Thailand Government Scholarship, SUT Graduate International Students Scholarship (Vithedbundit), and Thailand Science Research and Innovation (TSRI). In addition, we would like to express our gratitude to Associate Prof. Yupaporn Ruksakulpiwat and Assistant Prof. Piya-on Numpaisal for providing electrospinning instrument.

References

[1] Alven, S., Buyana, B., Feketshane, Z., & Aderibigbe, B. A. (2021).*Pharmaceutics, 13*(7). doi:10.3390/pharmaceutics13070964
[2] Casasola, R., Thomas, N. L., Trybala, A., & Georgiadou, S. (2014). *Polymer, 55*(18), 4728-4737. doi:10.1016/j.polymer.2014.06.032
[3] Dlugosz, M., Bulwan, M., Kania, G., Nowakowska, M., & Zapotoczny, S. (2012). *J Nanopart Res, 14*(12), 1313. doi:10.1007/s11051-012-1313-7
[4] Gu, Shu-Ying, & Ren, Jie. (2005). *Macromolecular Materials and Engineering, 290*(11), 1097-1105. doi:10.1002/mame.200500215
[5] Haider, Adnan, Haider, Sajjad, & Kang, Inn-Kyu. (2018). *Arabian Journal of Chemistry, 11*(8), 1165-1188. doi:10.1016/j.arabjc.2015.11.015
[6] Mahapatra, A., Garg, Nidhi, Nayak, B. P., Mishra, B. G., & Hota, G. (2012). *Journal of Applied Polymer Science, 124*(2), 1178-1185. doi:10.1002/app.35076

[7] Pan, Shu-Fang, Ke, Xiao-Xue, Wang, Ting-Yu, Liu, Qing, Zhong, Lu-Bin, & Zheng, Yu-Ming. (2018). *Industrial & Engineering Chemistry Research, 58*(2), 984-993. doi:10.1021/acs.iecr.8b04893

[8] Promnil, Siripanyo, Numpaisal, Piya-on, & Ruksakulpiwat, Yupaporn. (2021). *Materials Today: Proceedings, 47*, 3496-3499. doi:10.1016/j.matpr.2021.03.504

[9] Sadasivuni, Kishor Kumar, Rattan, Sunita, Waseem, Sadiya, Brahme, Snehal Kargirwar, Kondawar, Subhash B., Ghosh, S., Mazumdar, Payal. (2019). In *Polymer Nanocomposites in Biomedical Engineering* (pp. 331-373).

[10] Wang, Wei, Yu, Zhilong, Alsammarraie, Fouad K., Kong, Fanbin, Lin, Mengshi, & Mustapha, Azlin. (2020). *Food Hydrocolloids, 100*. doi:10.1016/j.foodhyd.2019.105411

Deciphering the discoloration in the manufacturing of natural rubber

J Sakdapipanich*

Department of Chemistry and Center of Excellence for Innovation in Chemistry, Faculty of Science, Mahidol University, Salaya campus, Phutthamonthon, Nakhon Pathom, 73170 Thailand

*Corresponding author: jitladda.sak@mahidol.ac.th

Abstract Many naturally occurring color components, especially non-rubber components, are present in natural rubber (NR) latex, limiting some NR applications, particularly for light-colored NR products in high demand. Non-rubber components, like lipids and proteins, can be removed from NR to reduce its yellow color. This study looked at the factors that influence the yellow index (YI) in NR and NR latex discoloration techniques. Proteins were a crucial factor directly determining the YI of NR. Thus, increasing the number of centrifugation washings, speed, and time was used to remove proteins to minimize the YI of NR. Percentage of total solid content (% TSC) of field NR latex, the blend ratio between cream rubber and skim latex, or bottom fraction from fresh-field NR latex collected from centrifugation process, were also important factors. The addition of sodium metabisulphites as a polyphenol oxidase (PPO)-reducing agent reduced the YI significantly. The oxidative breakdown of endogenous non-rubber components at high temperatures was also confirmed to serve the increase in YI value due to air-drying conditions of NR.

1. Introduction

Natural rubber (NR) is a major agricultural product of Thailand, produced from *Hevea brasiliensis* trees in both solid and concentrated latex form. Vehicle tires, automobile parts, dock fenders, gloves, condoms, and other consumer products use NR as a key raw material. Because NR is naturally yellowish, light-colored and colored-filled goods require light or extra-light graded rubber as a feedstock manufactured during the manufacturing process. In the case of concentrated NR latex, there is no specific grade for making light-colored latex products.

The yellow color of NR is influenced by clonal and seasonal changes, soil types, tapping frequency, and non-rubber components. The presence of carotenoids is thought to be responsible for the unique yellow hue of NR [1]. Carotenoids from RRIM600 and PB235 clones were extracted and analyzed, revealing the presence of four important carotenoids: lutein, zeaxanthin, carotene, and carotene [2]. Because polyphenol can react with compounds such as proteins *via* a non-enzymatic reaction, it has been identified as the most important component that causes discoloration of raw NR [3,4] and vulcanized NR [3]. Colorants extracted from various NR latex fractions, including fresh latex, bottom fraction, cream fraction, and Frey-Wyssling (FW) particles, contained similar constituents, such as carotenoids, tocotrienol esters, fatty alcohol esters, tocotrienols, unsaturated fatty acids, fatty alcohols, diglycerides, and monoglycerides, according to Sakdapipanich and colleagues [5.] Non-rubber coloring components such as polyphenol, proteins, and carotenoids, according to Madsa-ih and Cheewasedtham [6], impact the color of NR. Orthoquinones are produced by the oxygen-dependent oxidation of polyphenols catalyzed by polyphenol oxidases (PPO). Colorants are formed when these quinones react with amino acids and proteins found in latex [7]. PPO is a phytohormone that aids in plant defense on a tapping wound of the rubber tree has been found in both lutoid and FW particles, with the former having higher activity [4]. PPO is released from the ruptured particles, causing discoloration. When proteins and amino acids in NR latex combine with carbonyl molecules, including reducing sugars, aldehydes, and ketones in the presence of heat, they generate colorful substances known as the Millard reaction.

Many studies have been conducted to discolor NR latex by suppressing enzyme activity, removing the substrates involved through chemical treatment, or using both combinations [8]. It has been reported that stimulating the Hevea tree with an optimal dosage of ethephon suppresses enzymatic browning in NR [9]. To overcome coloring compounds inherent in NR, purifications by deproteinization [10] and transesterification [11] are procedures for eliminating proteins and lipids from NR latex/solution, respectively. Despite the fact that sodium metabisulphite (SMS) acts as an enzymatic browning inhibitor, it does not affect the color of NR [12]. As a result, this current work uses a colorimeter to determine a yellow index to figure out the main component that affects the yellowness of NR (YI). When scaled up, systematic research of the factor and the procedure to lower the YI will be conducted to implement a feasible way to manufacture light-colored NR and latex for industrial choice.

2. Experimental

Materials: Thai Rubber Latex Group Public Co., Ltd. provided fresh field latex (FL) from a plantation in Chonburi, Thailand, from *Hevea* rubber tree clone RRIM 600. FL-latex was treated with 0.6 percent w/v ammonia and 0.5% w/v potassium laurate, then filtered using a muslin cloth to eliminate contaminants and coagulum. The FL-latex was cast on a clean, dust-free petri dish and dried at room temperature to make fresh NR (FNR) film. Bangkok Synthetics Co. Ltd., Thailand, provided *cis*-1,4 polyisoprene or synthetic isoprene rubber (IR).

Sample preparation: Proteins were extracted from FL by incubating 30% DRC of FL with 0.04 % w/v proteolytic enzyme (Kao KP-3939) in the presence of 1 % w/v sodium dodecyl sulphate (SDS) for 16 h at 37°C [10], followed by washing twice with 12,000 rpm centrifugation for 30 min. Deproteinized NRL or DPNRL refers to the latex that has been obtained. The DPNRL was cast in a clean petri dish as a film and dried at room temperature. A reaction of a 0.5% w/v FNR solution in toluene with freshly produced sodium methoxide for 3 h at 37°C yielded transesterified NR (TENR) [11]. The same technique was used to make TENR following deproteinization (DP-TENR); however, DPNR was used as the starting material instead of TENR. FL-latex was centrifuged for 30 minutes at 25°C to produce concentrated NR latex (CNRL). The percentage total solid content (% TSC) of FL before centrifugation, centrifugation time/speed/number of washing cycles, and other aspects in the actual centrifugation process at the manufacturing that may affect the color of NR were also explored. The TSC of FL was changed from 15 to 40% before centrifugation at 12,000 rpm for 30 min. All of the CNRL that resulted were modified to 60% w/w TSC and 0.5% w/v potassium laurate (aq). The CNRL film was made from 4 mL of diluted CNRL (50 % w/w TSC) by casting it on a clean petri dish (60 mm diameter) and drying it at room temperature overnight, then drying it *under vacuum* until reaching a steady weight.

Characterization

Determination of color: According to ASTM E313 @ 10°, the yellow index (YI) of the rubber films was measured using a Hunter Lab (ColorQuest®XE) CIE colorimeter with an Illuminant D65 light source.

Determination of nitrogen content: A LECO FP 528 Nitrogen Analyzer was used to determine the nitrogen concentration of NR. The nitrogen content was determined by duplicate analysis with a standard deviation of 0.0015%, comparable to the EDTA standard.

Determination of ester content: Fourier-transform infrared (FT-IR) spectroscopy was used to analyze the ester content of each dry rubber sample by measuring the variety of mixture of methyl stearate as a standard and synthetic *cis*-1,4 polyisoprene, as described in our earlier work [11].

3. Results and discussion

Compared with the control FNR and synthetic IR rubber, Figure 1 illustrates the appearance and change in YI and N/ester contents of various types of NR, *i.e.*, DPNR, TENR, and TE-DPNR, after removing the non-rubber components (NRC). The YI of DPNR, TENR, and TE-DPNR all dropped from 24 for FNR to 10, 8, and 3 for DPNR, TENR, and TE-DPNR, respectively. The colors in Figure 1 correlate to the actual color of the rubber samples. FNR and DPNR were yellowish and pale yellow, respectively, whereas TENR was virtually colorless and TE-DPNR and IR were transparent. The N content, ester content, YI value, and color of TE-DPNR and IR were surprisingly equivalent. According to these findings, proteins seem to be the most important component dictating the yellowness of NR.

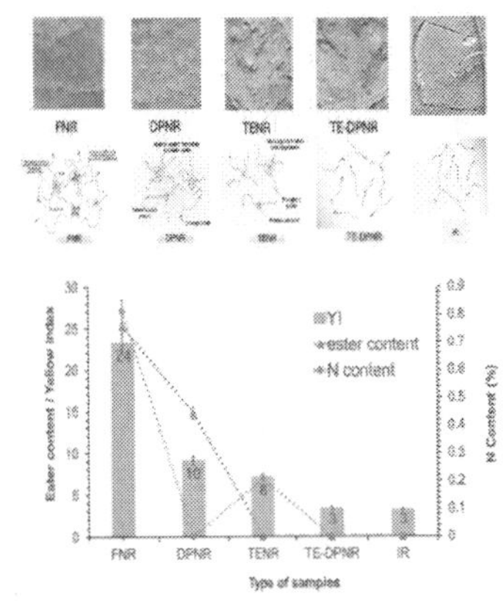

Figure 1. Total N content, ester content and YI of NR after various treatments, *i.e.*, DPNR, TENR, TE-DPNR, in comparison with FNR and synthetic IR.

In practice, centrifugation yields concentrated NRL from TSC of 30 to 60. As a result, numerous parameters were investigated thoroughly to decide the proper condition, including altering the percent TSC of FL, centrifugation time and speed, and the number of washings (as shown in Figure 2). The best speed is 12,000 rpm for 30 min, and the percent TSC should be less than 30%.

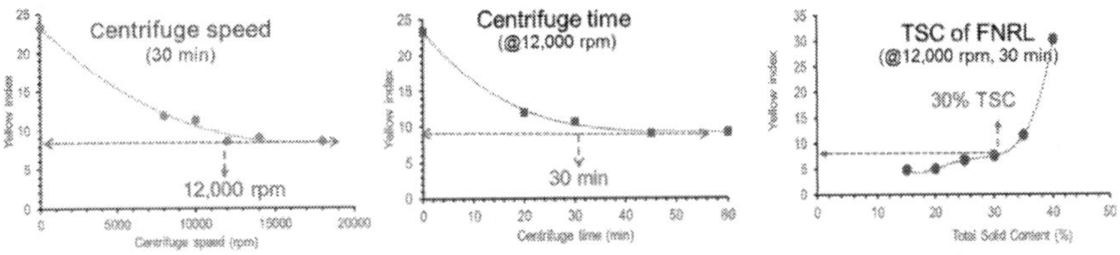

Figure 2. YI of CNRL before and after centrifugation at various conditions.

To investigate the component derived from the composition of FL after centrifugation, a mixture of cream rubber and skim latex and the bottom fraction were prepared. Figure 3 illustrates the change in ester and nitrogen contents as a function of YI values for various amounts of skim latex and bottom fraction mixed with the cream rubber. It is noticeable that as the nitrogen content of the skim latex and bottom fraction increased, the YI increased proportionally. The findings of (a) may be compared to the screw adjustment in the centrifugation at the factory, which will separate approximately 20-25%, as the so-called skim latex, which would produce the best CNRL quality. The N contented, and YI values increased dramatically as the BF content was increased in experiment (b), while the ester content remained constant. These findings suggest that removing as much BF before centrifugation will reduce the YI of the CNRL obtained.

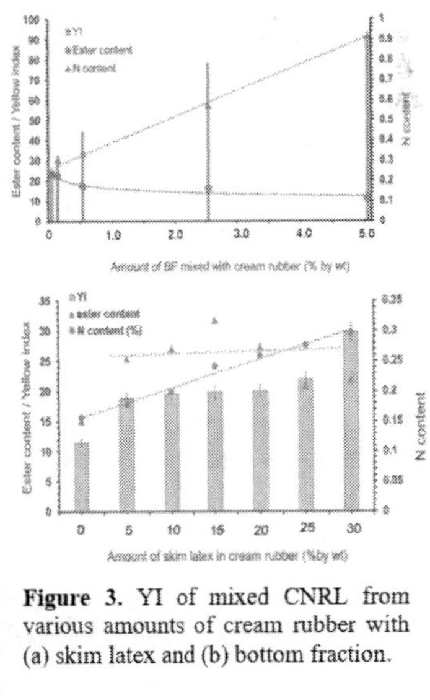

Figure 3. YI of mixed CNRL from various amounts of cream rubber with (a) skim latex and (b) bottom fraction.

Figure 4 shows how the YI of several NRs changed before and after sodium metabisulfite (SMS)was added. The incorporation of SMS as a PPO-reducer could improve the YI of rubber by approximately 10%. By comparison with the control NR dried at 50°C for 48 h and FNR, Figure 5 shows the change in YI and characteristics of NR after drying at various conditions. The darkening was seen in the samples 150_{1h}-NR and 150_{3h}-NR, which were dried at higher temperatures. Because of the oxidative degradation of the non-rubber components, this finding suggests that a high-temperature drying process could boost the high YI of NR.

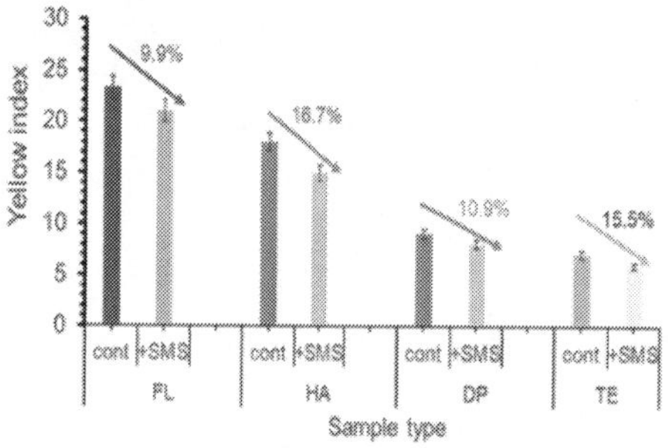

Figure 4. YI of various CNRLs mixed with SMS.

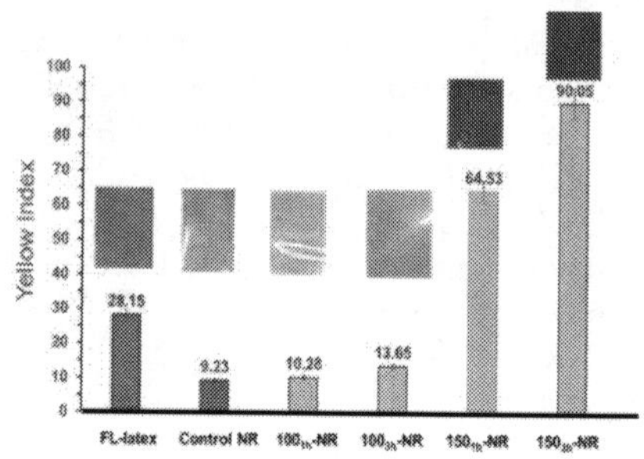

Figure 5. YI of CNRL dried at various conditions.

4. Conclusions

The niche CNRL and light-colored NR were established in this investigation by fine-tuning the centrifugation processes and removing non-rubber components, including proteins, and the drying process. The approaches are manageable and achievable for manufacturing consent with the smallest alteration to becoming a machine. A light color CNRL can meet the demand, and the higher processing charge can be ignored.

Acknowledgment

This research was funded by the Thailand Science Research and Innovation (formerly Thailand Research Fund) under Grant Number RDG62T0030, and the National Research Council of Thailand provided publication support (NRCT). The authors also appreciate Mahidol University and the Ministry of Higher Education, Science, Research and Innovation's Center of Excellence for Innovation in Chemistry (PERCH-CIC). The Thai Rubber Latex Group Public Co., Ltd. is honored cordially for supplying the NR latex.

References

[1] G. Verhaar, *Rubber World* 1965, **151**(5), 74.

[2] A. Bahari, Approaches to understanding diversity in rubber and carotenoid synthesis in *Hevea brasiliensis* latex 2019, Ph.D. Dissertation, University of Dundee.

[3] K. Komolpis, Effect of some non-rubber constituents on discoloration and cure characteristic of natural rubber 1991, M. Sc. Dissertation, Chulalongkorn University.

[4] D. Wititsuwannakul, N. Chareonthiphakorn, M. Pace, R. Wititsuwannakul, *Phytochem.* 2002, **61**, 115.

[5] J. Sakdapipanich, K. Insom, N. Phuphewkeaw, *Rubber Chem. Technol.* 2007, **80**, 212.

[6] Y. Madsa-ih, W. Cheewasedtham, *Eur. J. Sci. Res.* 2012, **90**(2), 255.

[7] N. Iyidogan, A. Bayindirh, *J. Food Eng.* 2004, **62**, 299.

[8] A. Mubarak, A. Ismun, S. B. A. Razak, M. M. Ariffin, *Malays. Appl. Biol.* 2018, **47**(4), 127.

[9] T. Chanwun, N. Muhamad, N. Chirapongsatonkul, N. Churngchow, AMB Express 2013, **3**, 14.

[10] J. Sakdapipanich, P. Rojruthai, In *Biotechnology – Molecular studies and novel applications for improved quality of human life*, Intech, Croatia. 2012, 213.

[11] J. Brzozowska – Hanower, P. Hanower, C. Lioret, *Physiol. Veg.* 1978, **16**, 231.

[12] R. Singthuean, S. Ratanawilai, P. Kaewpradit, *Proc. 5th PSU-UNS International Conference on Engineering and Technology (ICET-2011), Songkhla, Thailand*, 2011, 26.

IOP Conf. Series: Materials Science and Engineering 1234 (2022) 012011 doi:10.1088/1757-899X/1234/1/012011

Near infrared spectroscopy and machine learning classifier of crosslink density level of prevulcanized natural rubber latex

P Sirisomboon[1], **C H Lim**[2], **J Posom**[3,*]

[1]Department of Agricultural Engineering, School of Engineering, King Mongkut's Institute of Technology Ladkrabang, 1 Chalongkrung Road, Ladkrabang, Bangkok, 10520, Thailand
[2]Thai Rubber Latex Group Public Company Limited, Nongyai, Chonburi, 20190, Thailand
[3]Department of Agricultural Engineering, Faculty of Engineering, Khon Kaen University, Khon Kaen, 40002, Thailand

* Corresponding author: jetspo@kku.ac.th (Jetsada Posom)

Abstract. By toluene swell index for cross link density level of prevulcanized (PV) rubber latex knowledge, toluene swell of PV latex is measured for trading and production management. Therefore, aim of this research is to use the Fourier transform near infrared (FT-NIR) spectroscopy with machine learning to classify different cross link density levels by toluene swell index including, Unvulcanized (U) (> 160%swell), Lightly vulcanized (L) (100-160%swell), Moderately vulcanized (M) (80-100%swell), Fully vulcanized (F) (< 80%swell) of prevulcanized (PV) natural rubber latex of raw PV latex and 50% solids content PV latex (PV50). The result shows that toluene swell index of rubber prevulcanized latex could be 91.8% correct classified into L group and M group using PV50 MSC pretreated spectra with PLS-DA classifier. Unfortunately, sample obtained for this experiment were loss of U and F groups. In future, to develop the robust model, the sample of all crosslink density levels should be collected.

1. Introduction

NIR spectroscopy have been studied for feasibility determination of cross link density of rubber latex such as cross link density determined by prevulcanisate relaxed modulus (PRM) test [1] using NIR Fourier Transform spectrometer (FT-NIR) and [2] using low cost spectrometer (LCS) and toluene swell test [3] using FT-NIR, when reference methods has been used in rubber latex factory. For prevulcanized latex an effective model was developed using partial least squares regression with preprocessing (first derivative + straight line subtraction method). The coefficient of determination (R^2), root mean square error of cross validation of the calibration set were 0.85 and 2.98% and coefficient of determination (r^2), root mean square error of cross validation and bias of the validation set were 0.71, 3.93% and -0.005%, respectively. By PRM test for PV 50% (prevulcanised latex of 50% total solids content), FT-NIR gave coefficient of determination of validation (r^2) and root mean square error of prediction (RMSEP) of 0.88 and 6.74×104 Nm^{-2}, respectively, while scanned PV50 latex in quartz cup [1] and LCS gave 0.70 and 8.13×104 Nm-2, respectively, while dipped probe in PV50 latex in glass beaker [2]. For FT-NIR spectrometer to be used in prevulcanized latex production, Williams et al [4] indicated this model was for screening and some other approximate calibration (r^2 is between 0.66-0.81). Therefore, objective of this manuscript was to use FT-NIR spectroscopy and machine learning including partial least squares discriminant analysis (PLS-DA) and artificial neural network (ANN) classifier to differentiate cross link density levels including Unvulcanized (U), Lightly vulcanized (L), Moderately vulcanized (M), Fully vulcanized (F) of prevulcanized (PV) natural rubber latex of raw PV latex and 50% solids content PV latex (PV50). Table 1 shows NIR spectroscopy has pro and con for property evaluation.

Table 1. NIR technique and toluene swell technique pros and cons.

NIR technique	**toluene swell technique**
Fast (< 5 minutes)	Slow (2-3 h)
No chemical used, no sample preparation, no skilled analyst required	Make film. Use solvent toluene which is toxic
Equipment expensive to purchase	Tools used are simple (die cutter, petri dishes, roller).
Flexible as many constituents can be tested simultaneously	Can test only one parameter i.e., toluene swell
Separate calibration is needed for each parameter	No calibration needed
Lack of knowledge in its operation can affect its accuracy	Test is simple and reproducible

2. Experimental

2.1. Sample preparation

The PV natural rubber latex samples for the experiment were provided by the factory of Thai Rubber Latex Group Public Company Limited in the Nongyai district, Chonburi province, Thailand. Typical formulation of PV natural rubber latex includes 20% KOH, 20% Ammonium laurate, 50% Sulphur, 50% ZnO and 0% ZDBC of 0.07, 0.08, 0.30, 0.20 and 0.40 Phr (Parts per hundred rubber), respectively. Totally seven 100-metric tonne storage tanks of PV latices were selected over a year. Each week, 3 samples were drawn from each tank at different levels (upper, middle and lower) for 8 consecutive weeks starting from the date of production (totally 9 times). The solids content of PV latices varied from batch to batch between 60.0 - 61.5%. For the purpose of comparison, the total solids content of PV latices should be the same for every batch. As such, the PV latices of all the batches were diluted to 50% solids content, hence PV50.

2.2. NIR spectral acquisition

NIR spectral acquisition was followed [1] in Near Infrared Spectroscopy Research Center for Agricultural Product and Food (www.nirsresearch.com) at Department of Agricultural Engineering, School of Engineering, King Mongkut's Institute of Technology Ladkrabang, Bangkok, Thailand. The Fourier transform near infrared (FT-NIR) spectrometer (MPA, Bruker, Ettlingen, Germany) was used to scan the samples at the wavenumber range of 12,500-3600 cm^{-1}. The gold plate was scanned internally for background reading before scanning the samples. For spectra of PV and PV50 latices, they were scanned by cup of 43 mm in diameter and 50 mm in height with the base area of 1453 mm^2 made of quartz material contained latex sample scanned through quartz bottom with diffuse reflectance mode at room temperature of 25±2°C. The apparent absorbance values were recorded using Log 1/R.

2.3. Analysis of toluene swell by the reference method

The toluene swell method is normally used for measuring the crosslink density of compounded or PV latex and toluene swell product. Analysis of toluene swell followed [3]. The percentage toluene swell was calculated from equation (1).

$$\%\text{swell} = \frac{Y-X}{X} \times 100 \quad (1)$$

where X is initial diameter and Y is swollen diameter read from the graph paper. The variation in %swell with crosslink density, for conventionally cured natural rubber in toluene, may be broadly described (Table 2) [5]. Latex concentrate before the additional of compounding chemicals has no crosslink and the toluene swell will exceed 160 % and the latex film test sample may even dissolve in the toluene rendering it immeasurable.

Table 2. Percentage swell and crosslink density level.

%swelling	Crosslink density level
> 160%	Unvulcanized rubber
100-160%	Lightly vulcanized
80-100%	Moderately vulcanized
< 80%	Fully vulcanized

Figure 1 shows flow diagram for classification analysis. Detail of classification analysis is following:

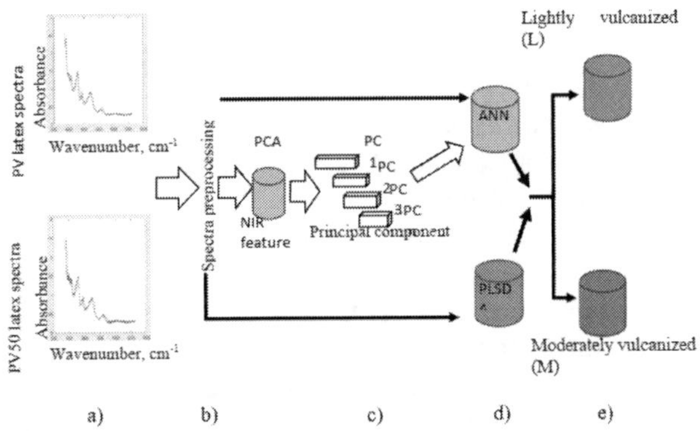

Figure 1. Flow diagram for classification analysis, a) spectra collection, b) spectra preprocessing, c) principal component score, d) modeling and e) classification performance determination.

Before the building of classification calibration models, principal component analysis (PCA) was used to detect outlier sample and no outlier was identified and removed. Before the calibration, by descendant running %swell data of 147 samples and assigning 3 samples in calibration set and 1 sample in prediction set (108 samples for developing a calibration equation and 39 of prediction set samples for validating the calibration equation). Absorbance spectra were pre-processed for optimal performance using multiplicative scatter correction and (MSC) standard normal variate (SNV). Principal component analysis (PCA) was performed on rubber latex spectra using maximum PC number determination on basis of error analysis described by Ni et al. [6] to get rid of risk to reserve noise and overfitting model if too many PC were used for modelling. PLS-DA and ANN were classifier for 4 levels of crosslink density of rubber latex using percentage toluene swell. However, only Lightly vulcanized (L) and Moderately vulcanized (M) levels were possible to be collected. The classification was performed on raw, pretreated spectra, and PCA pretreated spectra and the classification performance was assessed using sensitivity, specificity, and F1-score. Sensitivity indicated how accurate of classifier not to assign other group sample to be in model group. Specificity indicated how accurate of classifier to assign model sample to be in model group. F1 score indicated classifier either accurate of classifier not to assign other group sample to be in model group or accurate of classifier to assign model sample to be in model group. In our case, MATLAB R2021a (Mathworks Inc.) was used for spectral pretreatment, PCA, PLS-DA and ANN.

3. Results and discussion

Figure 2 shows a) PV and b) PV50 latex spectra scanned by FT-NIR spectrometer when water O-H bond vibration (1190, 1450 and 1940 nm) on 8285, 6888 and 5191 cm^{-1} (1207, 1452 and 1926 nm) were obviously illustrated and Para rubber polyisoprene absorbance band on 5924-5446 cm^{-1} (1688-1836 nm) and 4219 cm^{-1} (2370 nm) when compared to dried pure Para rubber sheet absorbance spectrum [7] where 1700-1884 nm absorbance band and 2278-2500 nm over absorbance band were illustrated.

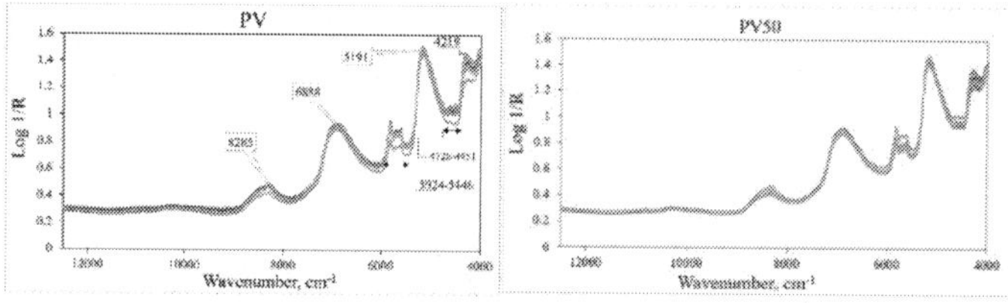

a b

Figure 2. a) PV and b) PV50 latex spectra scanned by FT-NIR spectrometer.

Statistical data of crosslink density of rubber latex samples indicated by dried rubber sheet % swelling in toluene where maximum, mean, minimum and standard deviation value were 120, 92.81, 82 and 8.58%, respectively. There was no sample of unvulcanized latex due to our sample was prevulcanized latex. Crosslinking starts immediately upon additional of the compounding chemicals. For even distribution of chemicals to ensure complete reaction, the latex compound has to be gentlely stirred for 24 h. As such, on the first measurement of the crosslink density, a certain amount of crosslinks had already occurred. By the 8th week, which was the upper boundary of our sample collection, most of the compounding chemicals would have been used up and the reaction continued at a very slow rate. However, upon further storage (about 3 months), where no sample collected, the toluene swell would reduce further to < 80% first by high modulus (HM), then middle modulus (MM) and low modulus (LM) as HM has a higher level of vulcanising chemicals. The cohesive mechanism of the polymer becomes stronger as the crosslinking increases, whereas the solvating power of the solvent remains constant. Therefore, a crosslinked high polymer substance is not dissolved but merely swelled, and the swelling decreases with increasing crosslinking. Thus, HM has more crosslinks than MM, and MM has more cross links than LM. The speed of crosslink formation was influenced by the external temperature of the environment.

Models from PV50 latex spectra was slightly better than from PV latex due to the effect of different total solids content in latex was suppressed. Among ANN, PCA-ANN and PLS-DA algorithm and different pretreatment methods (SNV and MSC), PLS-DA with MSC provided highest performance for classification with sensitivity, specificity and F1 score of prediction set of 87.5, 96.6 and 91.8% for lightly vulcanized level latex and 96.6, 87.5 and 91.8% for moderately vulcanized level latex, indicating PLS-DA classifier for lightly vulcanized level latex had 12.5% error in wrong assigning moderately vulcanized group and had 3.4% error in wrong assigning lightly vulcanized latex group and for moderately vulcanized level latex had 3.4% error in wrong assigning lightly vulcanized group and had 12.5% error in wrong assigning moderately vulcanized latex group and both classifier had 91.8% accuracy in classification. There was no literature reporting on classification of crosslink density of rubber latex. NIR spectroscopy exhibits good sensitivity offering detection limits on the order of 0.1% (w/w) concentration of constituent in quantitative analysis.[8] However, in qualitative analysis for group classification in case of vulcanized level by toluene swell of 4 groups (> 160%, 100-160%, 80-100%, < 80%) in Table 2 where range of percentage of any one group is wide. Therefore, it is possible to classify into different 4 groups. Group classify detection limit by NIR spectroscopy though has been not specified depends on how different of spectrum structure in different groups and how wide group reference properties for classification.

4. Conclusions

Crosslink density by toluene swell index of rubber prevulcanized latex could be 91.8% correct classified into lightly vulcanized latex group (100-160%swell) and moderately vulcanized latex group (80-100%swell) using FT-NIR PV50 MSC pretreated spectra with PLS-DA classifier. However, sample collected in this experiment were lack of unvulcanized rubber (> 160%swell) and fully vulcanized (< 80%swell). Therefore, more sample have to be collected to complete crosslink density range for develop applicable model to be used in prevulcanized rubber latex and product industries.

Acknowledgement

The authors would like to express their sincere thanks to Thailand Research Fund for the financial support (Grant number RDG5950011).

References

[1] Lim CH and Sirisomboon P 2017 J Near Infrared Spectrosc. 25 407–415.
[2] Lim CH and Sirisomboon P 2021 Industrial Crops and Product 159 1130162021.
[3] Lim CH and Sirisomboon P 2018 Journal of Near Infrared Spectroscopy 26(3) 159-168.
[4] Williams P, Antoniszn J and Manley M 2019 Near-infrared Technology: Getting the best out of light, AFRICAN SUN MeDIA, Stellenbosch, South Africa, 301 pages.
[5] Mausser RF 1987 (ed) The Vanderbilt latex handbook. 3rd ed. Norwalk, CT, USA, R.T. Vanderbilt Company Inc.
[6] Ni L-J, Zhang LG, Xie J, Luo JQ 2009 Analytica Chimica Acta. 633 43-50.
[7] Kaewkabthong A 2012 Master thesis of Faculty of Engineering, King Mongkut's Institute of Technology Ladkrabang, Bangkok, Thailand.
[8] Talebi M, Armstrong DW 2020 Water determination in Specification of Drug Substances and Products (Second Edition) Edited by: Christopher M. Riley, Thomas W. Rosanske and George Reid, Elsevier Ltd.

Study of nylon textile-reinforced natural rubber composite

N Choosang[1] and W Smitthipong[1,2,*]

[1]Specialized Center of Rubber and Polymer Materials in Agriculture and Industry (RPM), Department of Materials Science, Faculty of Science, Kasetsart University, Bangkok, 10900, Thailand
[2]Office of Research Integration on Target-Based Natural Rubber, National Research Council of Thailand (NRCT), Bangkok, 10900, Thailand

*Corresponding author: fsciwssm@ku.ac.th

Abstract. This study focuses on the production of nylon-reinforced natural rubber composites. In general, technical textiles serve as reinforcement and strength materials for a wide range of applications in rubber/textile composites. The adhesion between rubber and nylon is the most important factor affecting the assembly process and the strength of the finished product. The results showed that natural rubber reinforced with nylon textiles can be efficiently prepared by splicing a single layer of nylon fabric between two layers of rubber. The nylon textile-reinforced natural rubber composite was characterized by tensile testing machines and rubber curing characteristics, etc. The main result showed that the mechanical properties of rubber/nylon composites were higher than those of pure rubber. From the experimental results, it was found that nylon fabric can strengthen the natural rubber composite material for use with car tires.

1. Introduction

Natural rubber (NR) is a very helpful rubber which represents low hysteresis, high green strength, high resilience, and high tear properties, etc [1-2]. Nowadays, the use of synthetic fibre as reinforcement for polymer matrix composites has received a lot of interest [3]. The rubber application is used by some fibre-reinforcing materials inserted during the rubber layers, or the elastomer matrix is directly mixed with reinforcing fillers. The main objective of a reinforcing filler is to enhance the mechanical properties of the rubber composite [4], whilst fibre-based materials provide an additional goal of providing acceptable functional properties to the product [5]. The technical textile can be served as a reinforcing and strength-enhancer in the rubber/textile composite. The adhesion between rubber and textile is the most important factor in influencing the assembly process and finished product performance in rubber/textile composites [6]. The rubber matrix can be compounded with either reinforcing fillers or reinforcing fibre assembly. Technical textiles have received a lot of attentions and are widely used to apply to industrial applications such as tires, conveyors, and so on. In the fibre-reinforcing composite industry, the original of fibre are quite different, for example, glass fibre, carbon fibre, nylon, polyester, and aramide, etc [7].

Recently, the fabrics are applied to reinforce the elastomer in order to improve of the mechanical properties of the composite [8]. The objective of this work is to study the properties of reinforced nylon fabrics in NR composites, especially in the state of crosslinked rubber. Therefore, nylon fabric can be used as a reinforcing material for NR composite materials due to its high mechanical strength. It is expected that developed composite samples with nylon fabrics can be used to reinforce NR composites for use in tire applications.

2. Experimental

2.1 Materials

NR was STR20 grade (Standard Thai Rubber) by Sintongthai Rubber Co., Ltd., carbon black (N330) was supplied by Bridgestone Carbon black (Thailand) Co., Ltd., zinc oxide (ZnO) was supplied by Global chemical Co., Ltd., stearic acid was supplied by Chemmin Co., Ltd., sulphur was supplied by Utids Enterprise Co., Ltd., N-cyclohexyl-2-benzothiazole sulfenamide (CBS) was supplied by Shandong Sunsine Chemical Co., Ltd., and nylon textile had warp yarn of 20 lines/square inch and weft yarn of 3 lines/square inch coated with resorcinol formaldehyde latex (RFL) was supplied by Haiyang Technology Co., Ltd.

2.2 Preparation of NR sample

NR compound was prepared by a two-roll mill front rotor speed: rear rotor was 20:24.1 rpm at a temperature of 60°C (Shanghai Rubber-Machine, China). Firstly, the NR was masticated for 2 min to make the rubber softer. Then, the chemical ingredients were mixed in a conventional order, namely ZnO and stearic acid followed by carbon black (N330) and CBS before sulphur was added in the last chemical for the total mixing time about 24 min, the compound formulation is shown in Table 1. After cooling to room temperature, the rubber compounds were ready to prepare the composite.

Table 1. Compound formulation for the investigation in this study (parts per hundred of rubber: phr).

Ingredient	Rubber compound (phr)
STR20	100
N330	40
ZnO	3
Stearic acid	1
Sulphur	2
CBS	1

2.3 Preparation of NR/nylon composite

One layer of reinforcement nylon textile weight 2 grams was sandwiched between two rubber sheets (the rubber weight 25 grams per sheet). Then, the sandwich sample was compressed at 160°C for 4 min to receive a 1 mm thick composite sheet.

2.4 Characterizations of the Composite

The rubber composite was characterized by various techniques such as the curing characteristics of the NR compounds were determined at a temperature of 160°C following ISO 6502:1991 by using a Moving Die Rheometer (Gotech, M-3000A). This machine provides information on the curing behaviour of rubber compound.

The tensile tests were performed based on the ASTM D412 standard using a universal testing machine (Gotech, AI-7005) with a dumbbell type sample. The crosshead speed was 500 mm/min and a 500 N load cell was used.

The H-Test following ISO 4647 standard, using a universal testing machine (Gotech, AI-7005) was used to determine the adhesion between textile cord and vulcanized rubber. This method is applied by cords made from either natural or synthetic fibre.

3. Results and discussion

3.1 Curing characteristics of the rubber compound

The curing characteristics of NR compounds are shown in Table 2, it can be revealed the scorch time (ts2) and cure time (tc90) of the NR compound. The scorch time (ts2) generally defines the measure of the time at which vulcanization of the rubber compound begins and thus represents the time limit available for processing. The cure time (tc90) is the estimated time for 90% vulcanization of the rubber compound which should be an optimum time [2]. From the previous work, a rubber formulation exhibited scorch time at 1 min 17 sec and cure time at 2 min 22 sec due to almost twice amounts of zinc oxide (ZnO) and stearic acid compared to our formulation. The result of literature increased kinetic of vulcanization and then reduced scorch time (ts2) and cure time (tc90) almost twice compared to our result [9].

Table 2. Curing characteristic parameters of the rubber compound.

Formulation	ts2 (min)	tc90 (min)
Rubber compound	2	4

3.2 Mechanical properties of NR/nylon composite

The use of reinforced nylon textile in NR composites represent the improvement of mechanical properties of materials. Table 3 shows the tensile strength and elongation at break of crosslinked NR/nylon composite. We found that the tensile strength of NR/nylon composite is almost twice higher than that of NR alone form literature review, indicating the compatibility of nylon fabric with NR in the composite [1]. The nylon fabric better resists to the external force, thus it can reinforce the NR. The resorcinol formaldehyde latex (RFL) was used to coat on the surface of nylon fabric, the resorcinol can react with the nylon fabric whereas the latex can react with NR during vulcanization. So, the nylon fabric with RFL adheres well on the NR surface, becoming the good NR/nylon fabric composite. The H test method was used to assess the adhesion between rubber and fabrics. Therefore, the rubber industry prefers to offer nylon textiles to reinforce tire production.

Table 3. The mechanical Properties of rubber/nylon composite.

Sample	Tensile strength (MPa)	Elongation at break (%)	H-test (N)
Rubber/nylon composite	14.07	29.83	46.29

4. Conclusions

In this work, two layers of crosslinked NR was successfully reinforced by one layer of RFL-coated nylon fabric to become the NR/nylon sandwich composite. The results showed that the NR compound was prepared in a good manner and then exhibited the optimum cure time compared to the literature review. The nylon fabric with RFL adhesive adhered well on the NR compound and then became the NR/nylon composite after vulcanization. The H-test showed the compatibility at the interface between

NR and RFL-coated nylon fabric. The tensile strength of crosslinked NR/nylon composite was higher than that of crosslinked NR alone. For rubber/nylon woven composites, moreover, this research is useful technical information for applications in the composite industry.

Acknowledgement
This work was supported by Bangkok Metropolis Motor Company Limited in providing venues and research grants. We also would like to thank Specialized center of Rubber and Polymer Materials in Agriculture and Industry (RPM), Department of Materials Science, Faculty of Science, Kasetsart University, Bangkok, Thailand for capital for the analysis of test results.

References
[1] Smitthipong W, Suethao S, Shah D, and Vollrath F 2016 *Polym. Test.* **55** 17-24.
[2] Chollakup R, Suwanruji P, Tantatherdtam R, and Smitthipong W 2019 *J. Polym. Res.* **26** 37.
[3] Surajarusarn B, Hajjar-Garreau S, Schrodj G, Mougin K, and Amornsakchai T 2020 *Polym. Test.* **82** 106289.
[4] Chollakup R, Suethao S, Suwanruji P, Boonyarit J, and Smitthipong W 2021 *Comp. Adv. Mat.* **30** 1-6.
[5] Stone C, Windsor F, Munday M, and Durance I 2020 *Sci. Total Environ.* **718** 134689.
[6] Betts D, Sadeghian P, and Fam A 2021 *J. Build. Eng.* **38** 102184.
[7] Reowdecha M, Dittanet P, Sae-oui P, Loykulnant S, and Prapainainar P 2021 *Heliyon.* **7** 07176.
[8] Savic A, Antonijevic D, Jelic I, and Zakic D 2020 *Energy Build.* **229** 11051.
[9] Chung K, Hong Y, and Park M 2011 *Elastomers and Composites.* **46** 70-77.

Effect of potassium oleate on chemical structure-compression relationship of natural rubber foam

P Kunklang[1] and W Smitthipong[1,2,*]

[1] Specialized Center of Rubber and Polymer Materials in Agriculture and Industry (RPM), Department of Materials Science, Faculty of Science, Kasetsart University, Bangkok 10900, Thailand.
[2] Office of Research Integration on Target-Based Natural Rubber, National Research Council of Thailand (NRCT), Bangkok 10900, Thailand 10900.

*Corresponding author: fsciwssm@ku.th

Abstract. Rubber foam-based products, such as mattresses, pillows, and cushions, are becoming increasingly popular. Rubber foam's formulation is still being developed to improve the properties, because of its wide range of applications. Potassium oleate is a surfactant for producing the rubber foam, although it can be used in different concentrations to prepare the different properties of rubber foam. The Dunlop method was used in this work, due to its simplicity and easy-to-use in the foam preparation. The effect of potassium oleate, which was used as a foaming agent, was investigated in this study by reducing the amount of potassium oleate. The density, compression, and Fourier transform infrared (FTIR) spectroscopy results showed that both density and compression of foam samples with half amounts of potassium oleate are higher than those of control sample with normal amounts of potassium oleate. Moreover, the FTIR result revealed a significant change in spectra with the reductions of potassium oleate contents. Therefore, rubber foam formulations are still developed to provide the optimum properties for suitable applications.

1. Introduction

Rubber foam is utilized in a variety of products, such as mattresses, pillows, cushions and so on [1]. The Dunlop process is the most extensively used, due to its ease of formation, low production costs, and straightforward methodologies [2]. In addition to the production technique, the formulation has been considered very important factor because the foam formulation is a key factor to develop the suitable rubber foam.

In this study, potassium oleate (PO) as a surfactant is a foaming ingredient. PO was used as a comparing concentration because it is a foaming producing material. The technique of preparing the foaming process is the Dunlop process. Furthermore, the Dunlop process has been widely using in manufacturing, and potassium oleate is a foaming chemical that is easy to make [3], as more user-friendly. Supitta *et al* [3] investigated the relationship between morphology and elasticity of natural rubber foam which related to the concentration of the chemical blowing agent or PO. The results indicated that there are no significant differences in the macroscopic compression stress–strain curves between the control foam and foam with decreasing PO up to 45% samples, attributed to the fact that the rubber foam represented heterogeneous cell size distribution. When the concentration of the chemical blowing agent was decreased, the cell size decreased while the cell density increased. However, the porosity percentage was only slightly different. As a consequence, the mechanical properties of all samples were similar. The objective of this research was initiated to study the relationship between the chemical blowing agent concentration and the chemical structure-compression properties of natural rubber foam, especially at low concentration of PO.

2. Experiment

In the preparation of rubber foam, the chemicals are classified into four groups: the first group is the foaming agent or soap is potassium oleate (PO), the second group is the vulcanizing agent, such as ZDEC (Zinc diethyldithiocarbamate), ZMBT (Zinc-2 -Mercaptobenzothiazole), wingstay L, and sulfur, the third group is the initiator and gel stabilize as ZnO and DPG (Diphenylguanidine), and the last group is the gelling agent as SSF (Sodium silicofluoride). All the chemical was supplied by Thanodom technology company limited, Bangkok, Thailand. The formulation of rubber foam is represented in Table 1.

Before starting the mixing, the concentrated natural latex (CNL) was stirred to reduce ammonia (NH_3) with a blender for 1 min. After that, PO was added and stirred for 10 min. Then, ZDEC, ZMBT, Wingstay L, and sulfur were added and stirred for 1 min. Next, ZnO and DPG were added and stirred for 1 min. Then, SSF was added and stirred for 1 min. After that, foam sample was poured into a mold and leaved for 45 min before vulcanizing at 90°C for 1.45 h. Finally, the sample was washed and heated to 70°C for 4 h.

Table 1. The formulation of rubber foams in phr (parts per hundred of rubber)

Formula	Control sample	Half-soap sample
60% CNL	100	100
10% PO solution	3.30	1.65
50% Sulfur dispersion	2.00	2.00
50% ZDEC dispersion	1.00	1.00
50% ZMBT dispersion	1.00	1.00
50% Wingstay L dispersion	1.00	1.00
50% ZnO dispersion	2.80	2.80
33% DPG dispersion	0.56	0.56
12.5% SSF dispersion	1.84	1.84

The samples were measured for width, length, and height to calculate the volume. Then, the samples were weighed on a balance and calculated the density by using the weight and volume of the samples, calculated as the following [4]:

$$\text{Density} = \frac{M}{V} \qquad (1)$$

where M is the rubber foam sample's weight (kg) and V is the rubber foam sample's volume (m^3).
For Compression testing, the samples were determined by a texture analyzer (TA.XT Plus, Stable Micro Systems, Godalming, Surrey, UK). The samples were pressed down to 75% of the sample height at a speed of 0.1 m/s.

Attenuated Total Reflection-Fourier Transform Infrared (ATR-FTIR) spectroscopy (VERTEX 70, Bruker, Billerica, MA, USA) was used to analyse the chemical functional group of rubber foam with Ge crystal probe at 500–4000 cm^{-1}.

3. Results and discussion

First of all, Figure 1 shows the density and compression tests of foam samples. The density test results indicated that both foam samples represent the density almost in the same level, the half-soap sample tends to be higher density than the control sample. Because the number of gas bubbles on the foam rubber surface and the amount of potassium oleate decreased in which resulted in the possibility of an increase in condensate. However, decreasing the PO concentration by more than 50% results in a material becomes too dense and inappropriate with natural rubber foam products. Meanwhile, when PO was decreased, the compression strength of the half-soap sample was greater than that of the control sample at 17 kPa and 15 kPa, respectively [5].

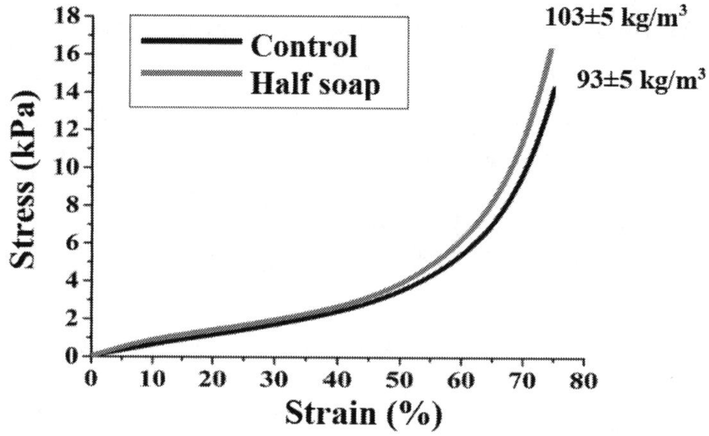

Figure 1. Results of the compression test showed the stress (σ)–strain (ε) curves of foam samples with varied concentrations of the chemical blowing agent (PO), and the densities of rubber foam at varying amounts of potassium oleate were also shown at the end of the line.

The ATR-FTIR results were studied in the chemical functional groups of both samples, as shown in Figure 2. The results of the control sample showed the peak at 3036-2850 cm^{-1} of C-H stretching (CH_3 and CH_2), the peak at 1663 cm^{-1} of C=C stretching in the double bond of natural rubber molecules [3], the peak at 836 cm^{-1} of C=C wagging and the peak at 1127-928 cm^{-1} of C-H stretching, respectively. We also found a peak at 1562 cm^{-1} which represented the asymmetric stretching modes of the carboxylate ion from potassium oleate [5]. The chemical structure of PO was shown in Figure 3. However, there are no significant differences of FTIR spectra between two samples.

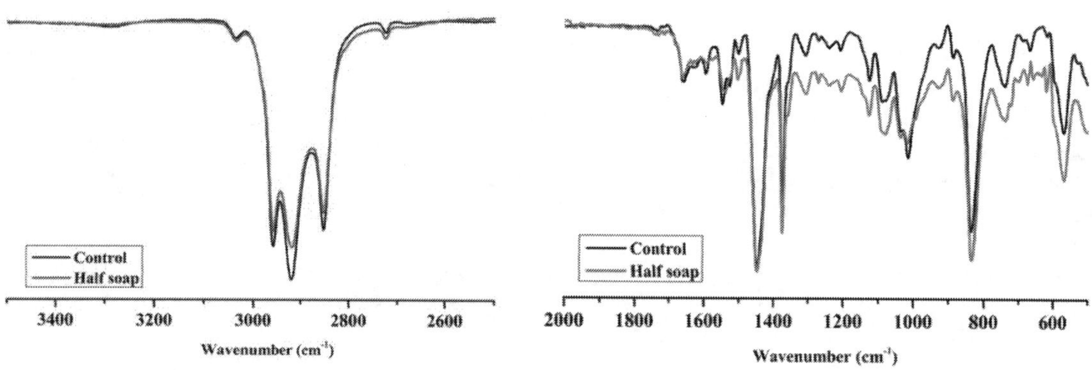

Figure 2. ATR–FTIR spectra of foam samples in transmittance from 500 to 3500 cm^{-1}.

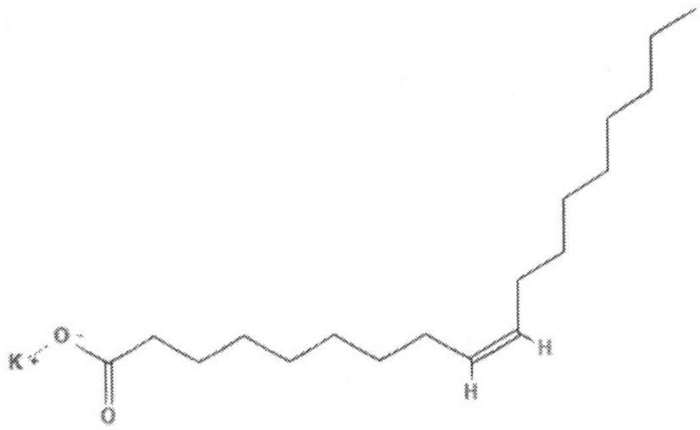

Figure 3. The PO structure with C=O bonds formed in the molecule [6].

4. Conclusions

In this research, effect of potassium oleate on chemical structure-compression relationship of natural rubber foam was studied. We found that foam density and compression increase with the decrease of PO concentration at 50%. When comparing both samples using ATR-FTIR, they exhibited similar results. So, we can adjust the foam density and mechanical properties of rubber foam using different PO concentrations.

Acknowledgment

Specialized center of Rubber and Polymer Materials for agriculture and industry (RPM), Faculty of Science, Kasetsart University, Thailand.

References

[1.] Suethao S., Ponloa, W., Phongphanphanee, S., Wong-ekkabut J, and Smitthipong W., 2021. Sci Rep. **11**(1). pp.1-12
[2.] Sirikulchaikij S., R Kokoo, and M Khangkhamano, 2020. Materials Letters, **260**. pp.1-15
[3.] Suethao S., Phongphanphanee S, Wong-ekkabut J, and Smitthipong W., 2021. Polymers, **13**(7). pp.1-14
[4.] Suksup R., Sun, Yu., Udomlak S., and Smitthipong W., 2019. J. Polym. Eng. **39**(4). pp.1-7
[5.] Popuri R., Amanapu H., Ranaweera C. K., Baradanahalli N. K., and Babu S. V., 2017. ECS J. Solid State Sci. Technol. **6**(12). pp.485-452
[6.] National Center for Biotechnology Information (2021). PubChem Compound Summary for CID 23665571, Potassium oleate. from https://pubchem.ncbi.nlm.nih.gov/compound/Potassium-oleate., Retrieved October 13, 2021

IOP Conf. Series: Materials Science and Engineering 1234 (2022) 012014 doi:10.1088/1757-899X/1234/1/012014

Study of non-negligible chemical reduction of ZnO in the rubber industry

W Wilaiwong[1] and W Smitthipong[1,2,*]

[1]Specialized Center of Rubber and Polymer Materials in Agriculture and Industry (RPM), Department of Materials Science, Faculty of Science, Kasetsart University, Bangkok, 10900, Thailand
[2]Office of Research Integration on Target-Based Natural Rubber, National Research Council of Thailand (NRCT), Bangkok, 10900, Thailand
*Corresponding author: fsciwssm@ku.ac.th

Abstract. Although the presence of small portion of zinc in the tire industry has been employed as activator, it should be concerned as the non-negligible chemical toxics and risks. The aim of this work was to study the effect of zinc oxide (ZnO) on physical and mechanical properties of tire rubber compound. The control characteristics of the composites were determined by the composites of practical compounds and vulcanizates by filling of sulfur and curing of a hot press machine. The properties of the composites, i.e. Mooney viscosity, curing time, tension, etc, were characterized. Generally, ZnO could work coupled with stearic acid (SA), the experimental was prepared by reducing ZnO at 20% and reducing stearic acid at approximately ratio to prevent excessive amount. The comparative data revealed the trend of the changes in the physical and mechanical properties just not lower at the same percentage. This recognized study supports the high possibilities in enhancing low ZnO and stearic acid content typical.

1. Introduction

Natural and synthetic rubber contains long polymeric chain with a high degree of flexibility and mobility since the molecules could be extensively movable and extendable in materials matrix excellently against applied load. Its high elasticity is most required property to be transformed to rubber compounding when have been modified into a network structure by crosslinking process. Sulfur crosslinking or so called 'vulcanization' is the most common and continued developing curing for unsaturated elastomers application, particularly in tire manufacturing industry. To gain desirable properties, i.e. high torque, high load aspect ratios, low rolling resistance, high wear resistance and good braking performance [1].

Zinc oxide has been applied as an activator supporting the vulcanization process among other rubber compounds to permit the 3D crosslinking network. ZnO, combined with SA, reduces the vulcanization time and improves rubber properties, even in the case of unaccelerated vulcanization [2]. The major industry for ZnO is rubber compounding, agriculture and ceramics, meanwhile zinc oxide is accounting for 59% of the zinc chemicals above zinc sulfate (36%) and zinc chloride (5%) [3]. The momentum of growing environmental awareness has become emphasized throughout toxicity of Zn possibly occurred in both acute and chronic forms [4]. The toxicological profile for Zinc [5] provided sample laboratory experiments indicated that a significant fraction of zinc may be released from tire rubber debris. S. Loranger *et al.* [6] collected zinc levels in snow on soil near an expressway in Montréal, Québec (Canada) at 15 m from expressway, the concentration of zinc in snow was 0.143 mg/L, while at150 m from the expressway, the mentioned concentration 0.029 mg/L. These conditions indicated a significant fraction of zinc dissolved in snow would be possibly released from tire rubber of vehicle using the expressway. A. Szwalec *et al.* [7] studied the impact of roads with various conditions of traffic and period of use on the distribution of Zn in the arable roadside topsoil. The concentrations of Zn, regardless of the distance from the road were 40.31 m/kg dry mattering topsoil of the road with 10 years of use and 18,000 vehicles/day and 63.97 mg/kg dry matter in topsoil of the road with 50 years of usage and 20,000 vehicles/day. The higher concentrations of Zn depended on traffic intensity, road type and distance from the road statistically were another evidence for non-negligible chemical loading by dust from the road.

Zinc compounds may be applied in the manufacturing and later be found as hazardous waste such zinc chloride, zinc oxide, zinc sulfate, and zinc sulfide. To protect workers, the Occupational Safety and Health Administration of the United States (OSHA) has set an average legal limit of 1 mg/m^3 for zinc chloride fumes and 5 mg/m^3 for zinc oxide (dusts and fumes) in workplace air during an 8-hour workday, 40-hour work week. Furthermore, a kind of community right-to-know regulations was issued to be guidance for industry's use and release of hazardous substances to prepare communities for and protect communities from chemical accidents. For instance, USA has implemented Emergency Plan and Community Right-to-know Acts (EPCRA) Section 313 for Reporting Guidance for Rubber and Plastics Manufacturing [8] particularly. This guidance states a sample step to identify, to transfer and to release the total amount of ZnO in process during the year determined by the annual throughput of the rubber mixture.

Several researches have been investigated over the decades in efforts to reduce ZnO, but further challenges of this complex chemical process are remaining significant in different solutions. For instance, the commercial grade of ZnO might be concerned about its low surface area, it further caused low ability to disperse the rubber matrix. [9] L. Pysklo *et al.* novel proposed novel such as active ZnO produced in a wet chemical process with smaller particle size and larger specific surface area to obtain the higher efficiency to be able to decrease ZnO loading. M. Guzmán *et al.* [10] proposed the use of nano-sized ZnO particles couple with more alternative metal oxides of Mg by the method consisted on the preparation of a polymer/metal salt complex. Despite previously explained novel or complex alternative, this study intended to provide the affordable green solution for reducing ZnO and reduce time to market as much as possible.

The main objective of this work was to investigate and adjust the rubber compounds to keep desired properties in the controlled compound from commercial sample could be. Basically, ZnO was used about 2-5 phr. Therefore, the middle of range was selected to be control sample in this study.

2. Experimental

2.1. Materials and preparations

Natural rubber (NR) type STR20 and butadiene rubber (BR) were used as the base materials for preparation of rubber composite with other substances following the formulation as shown in Table 1. Carbon black is fillers or reinforcement aids added to rubber formulations to meet application targets such as tensile strength and abrasion resistance 3 typical components for vulcanizing system are vulcanizing agents, activators and accelerator. This experimental used sulfur for vulcanizing, zinc oxide stearic acid for activating and then N-cyclohexyl-2-benzothiazyl sulfenamide (CBS) for accelerating. Another organic chemical was N-1,3-Dimethylbutyl-N'-phenyl-para-phenylenediamine (6PPD) used as an antiozonant [11].

Table 1. Compound formulation of rubber composites.

Ingredients	Recipe A (phr)	Recipe B (phr)
Natural rubber	50	50
Butadiene rubber	50	50
Carbon black (N550)	28	28
Aromatic oil	22	22
Zinc oxide	2.5	2
Stearic acid	1.1	0.9
Sulphur	1.4	1.4
CBS	1.4	1.4
6-PPD	4	4

All materials supplied from Thailand. The compounds were mixed in a two roll-mill mixer (Model R11-3FF; Kodaira Seisakusho; Japan) at a temperature of 70°C for 25 min by adding ingredients as following stage:

1. NR and BR 8 min.
2. ZnO and stearic acid 3 min
3. 6-PPD and carbon black and then aromatic oil 8.5 min.
4. CBS 1 min.
5. Sulphur 4.5 min.
6. Final Mix 2 min.

Each sample rubber compounds were then left 24 h prior to testing for which determine cure time (t_{90}) and scorch time. The rubber composites were molded into sheets with respective cure time, t_{90}, at 150°C using a hot press machine.

2.2 Characterization

The Mooney viscosity of the rubber compounds was determined using a Mooney viscometer in compliance with Standard ISO 289-1:2015. The cure characteristics were analyzed using a moving die rheometer (MDR) for 30 min at 50°C according to ASTM D5289-19. The vulcanized rubber sheet of each recipe was prepared 5 dumbbell shape samples with dimensions 115 mm x 25 mm x 2 mm. The mechanical properties were collected, according to ASTM D412-16, consisting of the tensile modulus, tensile strength and elongation at break.

3. Results and Discussion

Table 2. Properties of rubber composites.

Mechanical and Physical Properties	Recipe A	Recipe B
Mooney viscosity (+/- 5 a.u.)	31.60	34.60
t_{90} at 150°C (+/- 1 min)	17.61	16.64
Tensile Stress at Preset Point: Tensile strain 100% (MPa)	0.60	0.70
Tensile Stress at Preset Point: Tensile strain 300% (MPa)	2.10	2.10
Tensile Stress at Preset Point: Tensile strain 500% (MPa)	4.90	4.90
Tensile Stress at Break (MPa)	7.60	8.00
Elongation to Break (+/- 10%)	637.00	654.00

The experimental results of reducing ZnO and stearic acid in rubber compounds, that represented tire rubber component, could be determined in 2 groups of properties for exploring macrostructure of unvulcanized rubber compounds. Another important property of vulcanizates was mechanical characteristics.

Mooney viscosity is the resistance of uncured rubber to flow under stress depended on deforming temperature measured by rotational viscometer techniques. Recipe B, which mixed with less ZnO and stearic acid, took slightly higher viscosity than Recipe A within the same uncertainty measurement. It should be noted that fewer zinc in Recipe B could keep generating adequate rate of vulcanization growth.

According to the ration of amount of accelerator and amount of sulfur in this control recipe or Recipe A was equal to 1, the curing system could be defined as the semi-effective vulcanization (SEV) [12]. When the accelerator activities rise more than conventional system, the reactions took place more actively, the monosulfide crosslinks increase, and the degree of effective crosslinking of sulfur increases. Since couple of ZnO and stearic acid would form as zinc soap, then it further improved the solubility of ZnO in the rubber compound and with the accelerator to form a complex. This complex is very lively, reacting with sulfur to produce an extremely strong vulcanizing agent. As a results of Recipe B's t_{90} or curing time was almost the same time with that of Recipe A. The reduced of ZnO level in recipe B were not affect the curing time in this study.

Tensile stress of both Recipes at preset point of tensile strain 300%, which are frequently interested among researchers for determining effect of reinforcing with carbon black in rubber mattrix, was 2.1 MPa for both recipes. And also 100% and 500% point could be reported as same increasing from 0.6 to 4.9 MPa. So, there were no significant differences of mechanical properties between 2 recipes, even less ZnO of Recipe B.

4. Conclusion

This study focused on simply reduce zinc content in tire and to discover the effect of less activator on rubber compounds characteristics. The experimental recipe adopted from practical tire rubber ingredients with less 20% of ZnO. As results of viscosity, curing time and mechanical properties, those results were not change within the same uncertainty magnitude. It should be noted that amount of ZnO was adequate to activate reactions initially in this study. Anyway, further steps responded by ZnO and stearic acid together as zinc-soap to facilitate by increase solubility with accelerator to increase rate of sulfur crosslinking, the less amount of ZnO would affect the accelerating rate. However, the efficiency of reduced ZnO recipes were not as high as the control recipes. So, reduced ZnO recipe could be a good alternative to the control recipe when determine its properties at general working range for contributing to eco-friendly productions.

Acknowledgement

Specialized center of Rubber and Polymer Materials in Agriculture and Industry (RPM), Department of Materials Science, Faculty of Science, Kasetsart University, Bangkok, 10900, Thailand.

References

[1] Chollakup R, *et al.* 2021 *Adv. Compos. Mater.* **30** 1-6.
[2] Heideman G, *et al.* 2004 *Rubber Chem. Technol* **77**(3) 512-541.
[3] Zinc chemicals, Inorganic Chemical Economics Handbook Published April 2020, IHS Markit https://ihsmarkit.com/products/inorganic-zinc-chemical-economics-handbook.html
[4] Zinc's Fact Sheet for Health Professionals by the Office of Dietary Supplements (ODS) U.S. National Institutes of Health, Department of Health & Human Services available online: https://ods.od.nih.gov/factsheets/Zinc-HealthProfessional/ (access in September 2021)
[5] A Toxicological Profile for Zinc (August 2005), U.S. department of Health and human services , Public health service, Agency for Toxic Substances and Disease Registry (ATSDR).
[6] Loranger S, *et al.* 1996 *Environ. Pollut.* **92**(2) 203-11.
[7] Szwalec A, *et al.* 2020 *Environ Monit Assess* **192** 155.

[8] Emergency Plan and Community Right-to-know Acts (EPCRA) Section 313 for Reporting Guidance for Rubber and Plastics Manufacturing, U.S. Environmental Protection Agency Zinc
[9] Pysklo L, *et al.* 2007 *KGK rubberpoint* **60**(10) 548-553.
[10] Guzmán M, *et al.* 2011 *J. Appl. Polym. Sci.* **119**(4) 2048-2057.
[11] Brendan R, Walter W, The Science of Rubber Compounding, Science and Technology of Rubber (Third Edition), Academic Press, 2005, Pages 401-454
[12] Tang R, Dong W. Preparation and Characterization of Modified Natural Rubber Blends For Water Purification Treatment Usage. ChemRxiv. Cambridge: Cambridge Open Engage; 2020.

Study of particle size of natural latex/polysaccharide composite

K Wongphul[1], and W Smitthipong[1,2,*]

[1] Specialized Center of Rubber and Polymer Materials in Agriculture and Industry (RPM), Department of Materials Science, Faculty of Science, Kasetsart University, Bangkok, 10900, Thailand.
[2] Office of Research Integration on Target-Based Natural Rubber, National Research Council of Thailand (NRCT), Bangkok, 10900, Thailand.

*Corresponding author: fsciwssm@ku.ac.th

Abstract. The natural additive especially polysaccharide is interested in polymer composite field because it can be biocompatibility, biodegradability, non-toxic, low cost and easy to find. In addition, it should be used to reinforce the polymer matrix and to reduce the cost of material. This work focused on the comparison of particle size of natural latex with and without polysaccharide. Besides, the particle size of natural latex composite with polysaccharide was also investigated in various concentrations of polysaccharide. The natural latex with 5 wt.% of alginate represented the biggest particle size due to the agglomeration of rubber molecule with polysaccharide. The particle size of polymer latex is an important parameter to better understand the interaction between polymer and filler in the polymer composite.

1. Introduction

Generally, the concentrated natural latex (NL) is produced by fresh natural latex from *Hevea brasiliensis* tree. The fresh NL, in which is white colloidal liquid from the bark, is mostly composed of *cis*-1,4-polyisoprene (over 20,000 units) and few units of *trans*-1,4-polyisoprene. However, the dry natural rubber also consists of nonrubber component about 5-10% such as proteins, phospholipids, glycolipids, fatty acids, etc. In addition, the rubber chain of NL is the initiating and terminating ends, in which is proteins and phospholipids, respectively. The influence of nonrubber components can be presented the difference of NL properties. Furthermore, the effect of nonrubber component represents the high Mooney viscosity of natural rubber at long-time rubber storage due to the storage hardening phenomenon. The gel fraction of natural rubber is formed by proteins and phospholipids. Then, the natural rubber was numerous properties such as high tensile and tear strengths, low heat build-up, good processability, etc. Moreover, the natural rubber can be applied for many products such as pillow, gloves, flexible films and car tires [1-4].

Nowadays, the NL is developed by improving the properties of rubber, such as mechanical property, physical property, viscoelastic property, and so on. In addition, the polysaccharides are interested as natural additive, because it can be biocompatibility, biodegradability, non-toxic, easy to find and low cost [5]. Besides, the polysaccharides, as cellulose, chitin and starch, are used to reinforce the NL for improvement the mechanical properties of NL composite [6]. Moreover, the different of natural filler can be affected with the properties of materials, for example, xanthan gum was added to study the redox-responsive hydrogel with/without UV light irradiation with the carbonyl groups of polymer matrix [7].

Hyaluronate, alginate and pectate were also added in the synthetic polymer to study the responding of photoresponsive hydrogels with the carbonyl groups of polymer matrix [8]. So, the aim of this work is to study the particle size of NL and NL composite with alginate polysaccharide for understanding the interaction between polymer and filler in the polymer composite.

2. Experimental

2.1 Natural latex and natural latex composite with polysaccharide

First of all, concentrated natural latex (60 wt.% of dry rubber content or DRC, and 1.6 wt.% of nonrubber components) was supplied by Num Rubber & Latex Co. Ltd., Thailand. The NL was diluted into 20 wt.% of DRC. After that, the NL (20 wt.% DRC) was stirred for 15 min with the solution of low viscosity alginic acid sodium salt (Alg) from brown algae, in which was purchased from Sigma-Aldrich, in various concentrations: 1, 3 and 5 wt.%, respectively.

Table 1. The formulations of natural latex (NL) and natural latex composite with polysaccharide.

Formulas	NL/0Alg	NL/1Alg	NL/3Alg	NL/5Alg
NL (wt.% DRC)	20	20	20	20
Alg (wt.%)	-	1	3	5

2.2 Characterization

The natural latex and natural latex composite with alginate were analysed the particle size with Zeta sizer (Ultra/Pro, Malvern Panalytical Ltd.). The samples were diluted using 0.01 g of NL from Table 1 in 10 g of deionized water (from Scientific Equipment Center, Kasetsart University, Thailand) and stirred them until well mixing. Finally, the diluted sample 1 mL was transferred into 10 g of deionized water for the Zeta sizer characterization.

3. Results and Discussion

Four types of natural latex (NL) sample with/without alginate (Alg) were prepared: NL/0Alg, NL/1Alg, NL/3Alg and NL/5Alg. We found that the peak of the particle size of the natural latex with/without alginate (Alg) presented bimodal distribution (Figure 1), as small and large particles of NL sizing. This relates to the general structure of the natural rubber (*Hevea brasiliensis*) with nonrubber components. The structure of the natural rubber composes of rubber hydrocarbon (main chain) and nonrubber, especially the proteins and phospholipids in both serum and end chains. In addition, the small particles presented the nonrubber component interaction, while the large particles exhibited rubber hydrocarbon. The particle sizes of rubber hydrocarbon chain and nonrubber components were 0.4-0.75 μm and 0.08-0.2 μm, respectively [1,3,9]. Figure 1 also showed the volume of small and large particles size of NL with Alg from 0 to 3 wt.% was approximately 1-4% and 12-13%, respectively. However, NL/5Alg represented the highest volume (19%) of big particle size, indicating long chain of alginate can be formed agglomeration with rubber molecule.

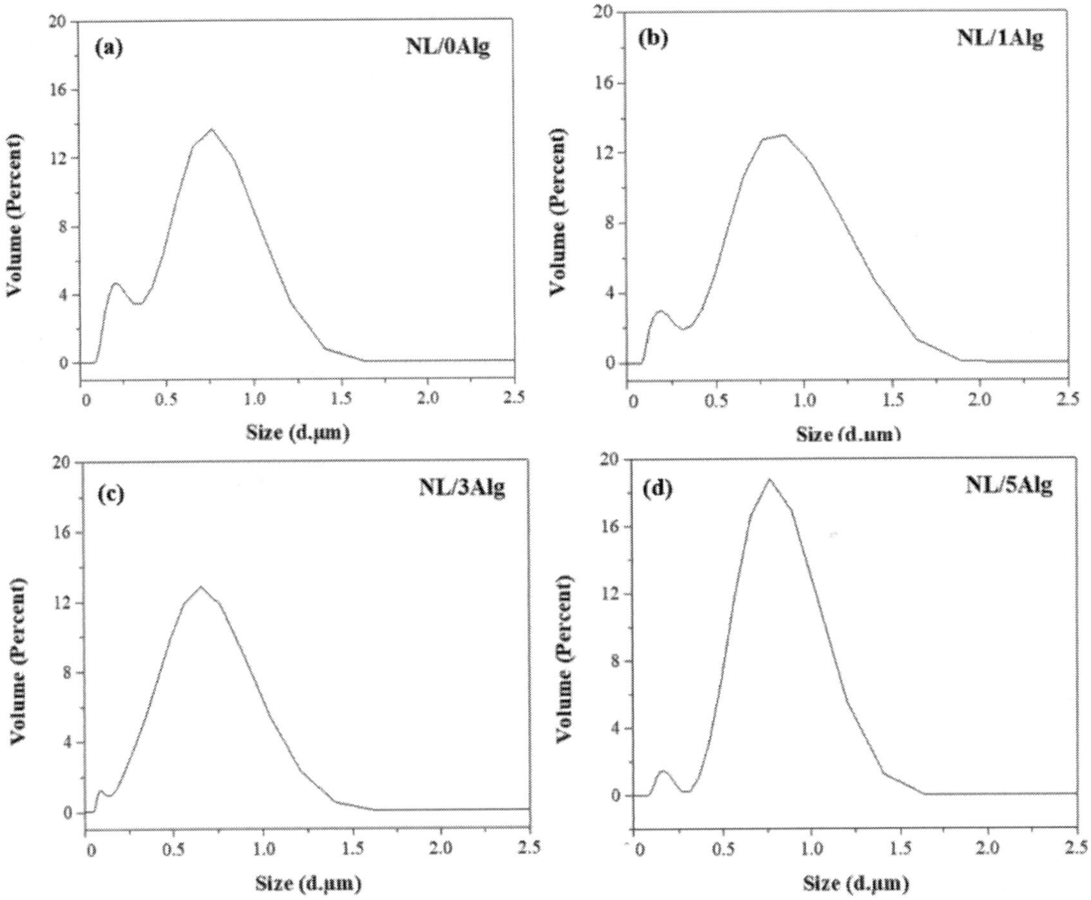

Figure 1. Particle size distribution of samples from the Zeta sizer: (a) natural latex without polysaccharide (NL/0Alg), (b) natural latex composite with 1 wt.% of polysaccharide (NL/1Alg), (c) natural latex composite with 3 wt.% of polysaccharide (NL/3Alg), and (d) natural latex composite with 5 wt.% of polysaccharide (NL/5Alg).

Table 2. Particle size of NL and NL composite with polysaccharide from the Zeta sizer.

Sample	Diameter of particle size (µm)	
	1st distribution (small particles)	2nd distribution (large particles)
NL/0Alg	0.20 ± 0.05	0.79 ± 0.26
NL/1Alg	0.19 ± 0.03	0.84 ± 0.07
NL/3Alg	0.18 ± 0.05	0.84 ± 0.22
NL/5Alg	0.20 ± 0.05	0.95 ± 0.18

Table 2 showed the particle size sample with/without alginate (Alg) in small and large particles. For the small particle size, all the samples exhibited almost the same size. Whereas the large particle size of sample with/without Alg until 3 wt.% showed also the same size, except the NL/5Alg possessed a trend to increase the particle size which may come from the agglomeration between rubber molecule and Alg [4]. So, the concentration of polysaccharide could affect the particle size of natural latex. The homogenous particle size can be applied for the thin film products.

4. Conclusions

In this research, the particle size of natural latex and natural rubber latex composite with polysaccharide represented bimodal distribution. The small particle size was nonrubber components, while the large particle size was main chain (rubber hydrocarbon). The NL/5Alg showed the highest volume and large particle size, certainly due to the agglomeration of alginate chain with rubber molecules. Thus, the concentration of Alg affects the particle size of natural latex. The applications of NL/Alg are rubber, glove and condom with natural filler as alginate to improve the mechanical properties.

Acknowledgement

Specialized center of Rubber and Polymer Materials in Agriculture and Industry (RPM), Department of Materials Science, Faculty of Science, Kasetsart University, Bangkok, 10900, Thailand.

References

[1] Chollakup R, Suwanruji P, Tantatherdtam R and Smitthipong W 2019 *J. Rubber Res.* **26** 37.
[2] Yunyongwattanakorn J, Tanaka Y, Kawahara S, Klinklai W and Sakdapipanich J 2003 *Rubber Chem. Technol.* **76** 1228–1240.
[3] Rojruthai P, Thawiwan K and Sakdapipanich J 2020 *J. Rubber Res.* **23** 353–364.
[4] https://docplayer.net/21048858-Rubber-chemistry-matador-rubber-s-r-o.html; October 2021.
[5] Kang M, Oderinde O, Liu S, Huang Q, Ma W, Yao F and Fu G 2019 *Carbohydr. Polym* **203** 139-147.
[6] Thomas K. S, Parameswaranpillai J, Krishnasamy S, Beguma P.M.S, Nandi D, Siengchin S, George J J, Hameed N, Salim Nisa.V. and Sienkiewicz N 2021 *Carbohydr. Polym.* **2**.
[7] Kang M, Oderinde O, Liu S, Huang Q, Ma W, Yao F and Fu G 2019 *Carbohydr. Polym* **203** 139-147.
[8] Giammanco G E. and Ostrowski A. D. 2015 *Chem. Mater.* **27** 14 4922–4925.
[9] Wei Y, Zhang H, Wu L, Jin L and Liao S 2017 *MOJ polym. sci.* **1** 197–199.

Study of yield percentage of epoxidized natural rubber preparation

K Panmanee[1] and W Smitthipong[1,2,*]

[1] Specialized Center of Rubber and Polymer Materials in Agriculture and Industry (RPM), Department of Materials Science, Faculty of Science, Kasetsart University, Bangkok, 10900, Thailand.
[2] Office of Research Integration on Target-Based Natural Rubber, National Research Council of Thailand (NRCT), Bangkok, 10900, Thailand.

*Corresponding author: fsciwssm@ku.ac.th

Abstract. Natural rubber (NR) mostly composes of *cis*-1,4-polyisoprene as rubber part and a few amounts of nonrubber part, mainly proteins and lipids. For the modification of NR, hydrogen peroxide and formic acid can be applied to produce epoxidized natural rubber (ENR) in the latex stage. The oxygen atom is added to the double bond of NR during epoxidation, resulting in oxirane (epoxide). The ENR reaction was allowed to operate at 60°C for 4 h. Characteristic infrared peaks at 870 and 835 cm^{-1} from Fourier Transform Infrared (FTIR) spectroscopy were used to calculate the mole percent of epoxidation. When the reaction time was increased, the proton-nuclear magnetic resonance (^{1}H-NMR) spectra of ENR showed an increase in mole percent of epoxidation. So, FTIR and NMR could be used to characterize the yield percentage of ENR preparation.

1. Introduction

Nowadays, NR, *cis*-1,4-polyisoprene, is significant and widely used in many fields [1-4]. NR in the latex stage is a colloid of milky white, this type of latex also includes nonrubber components such as lipids, proteins, minerals, etc [3]. As NR is a nonpolar material, it cannot react with polar compounds and must be converted to ENR. The epoxide group can increase the polarity of NR. During epoxidation process, the oxygen atom is reacted to the double bond of NR, resulting in the oxirane (epoxide). More polar functional groups should be combined with polar rubber for more applications. Many modification methods of NR are studied, but ENR is one of the most interesting and straightforward process. When formic acid and hydrogen peroxide are reacted to natural latex, it can be converted to epoxidized natural rubber (ENR) [5]. The ENR has many epoxide groups that can react to other polar components such as silica, even amine molecule, etc [6].

Recently, ENR can be reacted by polydopamine (PDA) and then crosslinked by PDA-metal coordination network. The obtained ENR with PDA-metal network represented mechanical properties with self-healing capability [7]. The objective of this work is to convert NR to ENR and then estimate the epoxidation percentage of NR at various reaction times. This can compare mole percent of epoxidation from absorbance of Fourier Transform Infrared spectroscopy (FTIR), deconvolution of FTIR peak, and

^{1}H-NMR peak. However, the ^{1}H-NMR technique is more accurate than the FTIR technique, but it takes longer and is more complex.

2. Experimental

Concentrated natural latex (60 wt.% of dry rubber content or DRC) was supplied by Num Rubber & Latex Co., Ltd., Trang, Thailand. Igepal CO-890 was purchased from Sigma Aldrich. Formic acid, hydrogen peroxide, and methanol were purchased from Merck.

Deionized water was used to dilute the concentrated natural latex to 20 wt.% DRC. After that, Igepal CO-890 was added to latex and then stirred for 4 h to remove ammonia. Next, formic acid and hydrogen peroxide were added to the latex in Table 1. The reaction was processed and stirred for 4 h at 60°C. The sediment was precipitated with methanol, washed with deionized water, and dried for 24 h at 50°C in a hot air oven.

Table 1. Recipe of ENR preparation.

Formula	**Weight (g)**
Concentrated natural latex (60 wt.%)	108.36
Igepal CO-890	22.75
Formic acid	29.25
Hydrogen peroxide	55.25

For the characterization of rubber, ^{1}H-NMR spectra were obtained by Bruker spectra operating at 400 MHz (AVANCE III HD Nanobay). The chemical shift of reference to the signal is 7.26 ppm. The sediment (5 mg) of rubber was dissolved by chloroform-d ($CDCl_3$) solvent in an NMR tube. The ^{1}H-NMR spectra can be used to calculate epoxidation percentage using the following equation (1):

$$\%\ \text{Epoxidation} = \frac{I_{2.7}}{I_{2.7} + I_{5.1}} \times 100 \qquad (1)$$

where $I_{2.7}$ and $I_{5.1}$ represent the integral peak area of chemical shift of proton of epoxide group and unsaturated carbon (C=C) proton, respectively.

The IR spectra were measured using Attenuated Total Reflection-Fourier Transform Infrared (ATR-FTIR) spectroscopy (VERTEX 70, Bruker, Billerica, MA, USA). The ENR in the latex stage was dropped in methanol, washed with deionized water, and dried for 24 h at 50°C in a hot air oven. The sample was measured between 400 and 4000 cm^{-1} at 2 cm^{-1} resolutions using 64 scans to characterize the functional group of rubber. We can calculate epoxidation percentage using the following equation (2):

$$\%\ \text{Epoxidation} = \frac{a_{870}}{a_{870} + a_{835}} \times 100 \qquad (2)$$

where a_{870} and a_{835} represent the absorbance of -C-O-C- (epoxide) and C=C-H (*cis*-1,4-polyisoprene), respectively.

The integral peak area of FTIR was used to calculate the deconvolution method (OriginPro 9.0, OriginLab), which can be used to analyze data from FTIR results by selecting peaks at 870 and 835 cm^{-1} and then adjust the baseline of the peak. However, we can calculate epoxidation percentage using the following equation (3):

$$\% \text{ Epoxidation} = \frac{A_{870}}{A_{870} + A_{835}} \text{ X } 100 \qquad (3)$$

where A_{870} and A_{835} represent the integral peak area of -C-O-C- (epoxide) and C=C-H (*cis*-1,4-polyisoprene), respectively.

3. Results and discussion

The sample was prepared by precipitation the latex with methanol, washed with deionized water, and dried at 50°C for 24 h to remove impurities. ATR-FTIR spectra of NR and ENR at different reaction times for 1 h, 2 h, 3 h, and 4 h are shown in Figure 1. The ATR-FTIR peaks of NR was determined at 2960, 2920, 1660, 1450, 1370, and 835 cm^{-1} which corresponded to C-H stretching of CH_3, C-H stretching of CH_2, C=C stretching, C-H bending of CH_2, C-H bending of CH_3, and C=C-H (*cis*-1,4-polyisoprene), respectively. The double bond of the NR backbone was converted to epoxide group and found new peaks at 1240 and 870 cm^{-1} which related to functional group of -C-O-C- (epoxide). When reaction time increased, the functional group of -C-O-C- (epoxide) at wavenumber 870 cm^{-1} was increased. So, mole percent of epoxidation can be calculated from 870 and 835 cm^{-1} which related to -C-O-C- (epoxide) and C=C-H (*cis*-1,4-polyisoprene), respectively (Table 2).

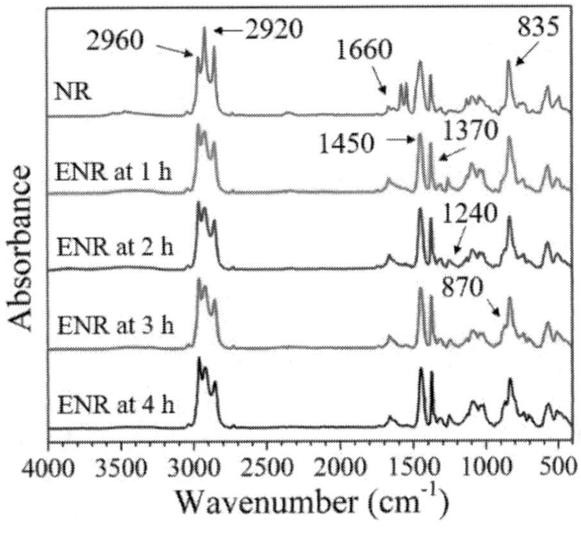

Figure 1. ATR-FTIR spectra of NR and ENR at different reaction times for 1 h, 2 h, 3 h, and 4 h.

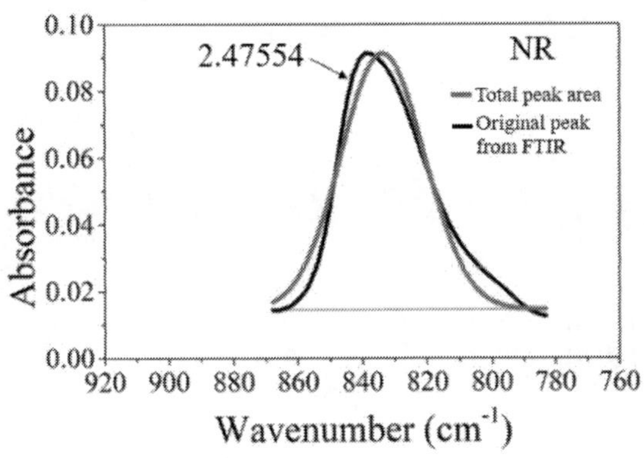

IOP Conf. Series: Materials Science and Engineering 1234 (2022) 012016 doi:10.1088/1757-899X/1234/1/012016

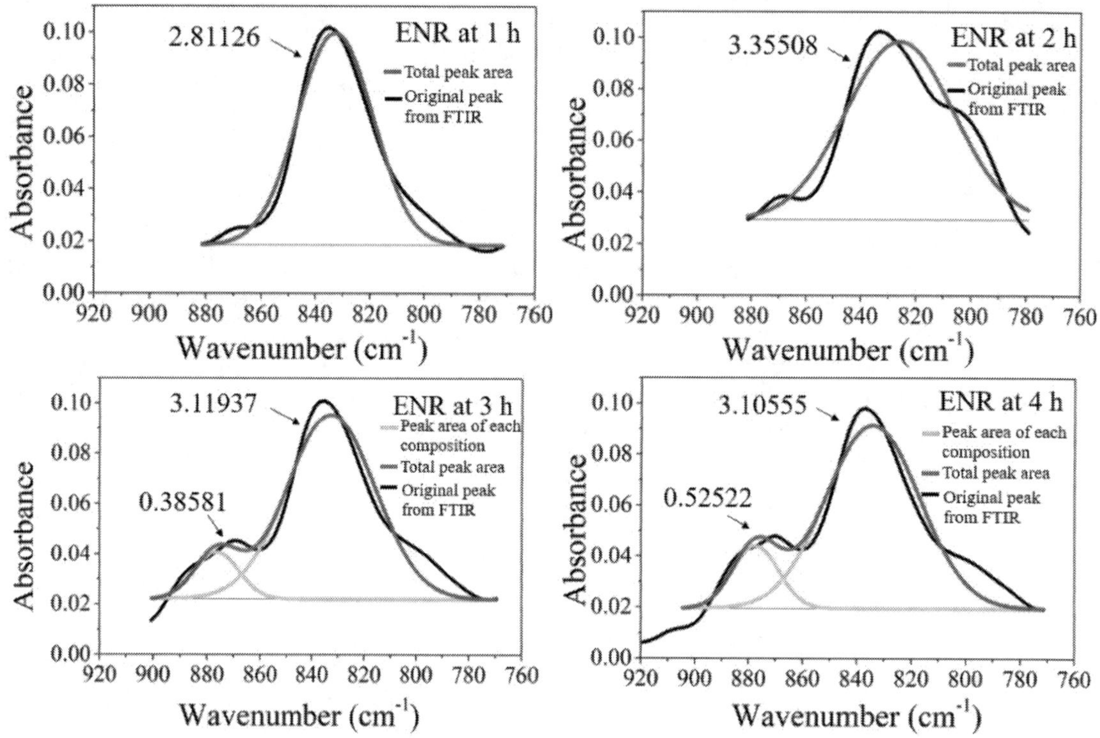

Figure 2. ATR-FTIR spectra of NR and ENR from deconvolution method at reaction times for 1 h, 2 h, 3 h and 4 h.

The absorbance and deconvolution method from FTIR spectra were applied for two peaks at 870 and 835 cm^{-1}. For the reaction times of 3 and 4 h, the approach of deconvolution method can be calculated the mole percent of epoxidation. The mole percent of epoxidation for 3 and 4 h calculated from FTIR spectra (Table 2) was 30.98 and 34.83 percent, respectively. As shown in Figure 2, FTIR peaks with deconvolution method cannot be calculated the mole percent of epoxidation for 1 and 2 h. The obtained mole percent of epoxidation for 3 and 4 h from deconvolution method was 11.01 and 14.47, respectively. The mole percent of epoxidation from absorbance was higher than that from deconvolution method.

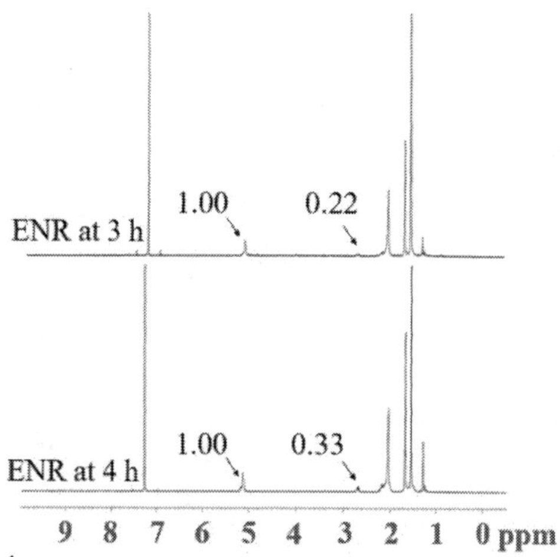

Figure 3. ^{1}H-NMR spectra ENR of reaction time at 3 h and 4 h.

Table 2. Comparison of %Epoxidation from FTIR absorbance, deconvolution method of FTIR peaks, and ^{1}H-NMR spectra.

Formula	%Epoxidation calculated from FTIR absorbance	%Epoxidation calculated from deconvolution method of FTIR peaks	%Epoxidation calculated from ^{1}H-NMR
NR	0	0	0
ENR at 3 h	30.98	11.01	18.03
ENR at 4 h	34.83	14.47	24.81

The ^{1}H-NMR spectra presented chemical shift of proton at 2.7 and 5.1 ppm which corresponded to epoxide group and unsaturated carbon proton, respectively (Figure 3). We found that the mole percent of epoxidation calculated from NMR for 3 and 4 h was 18.03 and 24.81%, respectively (Table 2). The results obtained from NMR was mostly in the middle between those from FTIR absorbance and deconvolution method. Generally, the ^{1}H-NMR was more accurate than calculating the FTIR absorbance peak and deconvolution method, because the ^{1}H-NMR directly analysed the connecting atoms to proton. On the other hand, the calculation from FTIR absorbance peak presented the highest mole percent of epoxidation because of the combination of other chemical functional peaks.

4. Conclusions

In this research, the rubber samples were prepared by precipitation in methanol, washed by deionized water, and dried at 50°C for 24 h to remove impurities. We prepared the ENR at different reaction times for 3 and 4 h which estimated the yield results from the FTIR absorbance method, the deconvolution method of FTIR peaks, and ^{1}H-NMR. The mole percent of epoxidation was calculated using FTIR spectra at 870 and 835 cm^{-1} and the ^{1}H-NMR at 2.7 and 5.1 ppm which corresponded to epoxide group and unsaturated carbon proton, respectively. When the reaction was increased, the mole percentage of epoxidation increased for these three methods. Furthermore, the obtained values from NMR were almost in the middle values between two methods from FTIR. However, the ^{1}H-NMR is a complicated method and takes long time to obtain the results. Whereas the FTIR technique is easy to operate and can estimate the results very fast. It should be noted that the ^{1}H-NMR is more famous than the FTIR for calculation the mole percentage of epoxidation of natural rubber [6].

Acknowledgement

Specialized center of Rubber and Polymer Materials in agriculture and industry (RPM), Department of Materials Science, Faculty of Science, Kasetsart University, Bangkok, 10900, Thailand.

References

[1] Suethao S, Ponloa W, Phongphanphanee S, Wong-Ekkabut J and Smitthipong W 2021 *Sci. Rep.* **11** 1.
[2] Chollakup R, Suethao S, Suwanruji P, Boonyarit J and Smitthipong W (2021) *Adv. Compos. Mater.* **30** 1-6.
[3] Smitthipong W, Punsuvon V, Janchai P and Vaithanomsat P 2017 *Rubber Chem. Technol.* **90** 4.
[4] Smitthipong W, Nardin M, Schultz J, Nipithakul T and Suchiva K 2004 *J Adhes Sci Technol* **18** 12.
[5] Chuayjujit S, Nutchapong T, Saravari O and Boonmahitthisud A 2015 *J. Met. Mater.Miner.* **25** 1.
[6] Cao L, Gong Z, Liu C, Fan J and Chen Y 2021 *Compos. Sci. Technol.* **207** 108750.
[7] Han Y, Wu X, Zhang X and Lu C 2017 *ACS Appl. Mater. Interfaces* **9** 23.

Evaluation on physicochemical properties and stability of Thanaka bark powder for natural face powder products

N Y Hnin[1] and M Kanlayavattanakul[1,2*]

[1] School of Cosmetic Science, Mae Fah Luang University, Chiang Rai 57100, Thailand
[2] Phytocosmetics and Cosmeceuticals Research Group, Mae Fah Luang University, Chiang Rai 57100, Thailand

*Corresponding author: mayuree@mfu.ac.th

Abstract. Thanaka (*Hesperethusa crenulata* Roem.) has been traditionally applied as cosmetics in Myanmar and the northern of Thailand [1] and exhibited anti-oxidant, anti-inflammatory, anti-microbial, skin lightening, and sun protection properties [1-4]. Thanaka's specifications for cosmetic development are sparse to be presented, although its flowability was reported [1]. Of which, the bark was more suitable than the stem for solid dosage form development [1]. Thus, this study was aimed to assess the physicochemical properties and stability of Thanaka bark powder. The yellowish light brown powder ($L^* = 75.63 \pm 0.01$, $a^* = 5.58 \pm 0.02$ and $b^* = 23.45 \pm 0.02$) (Figure 1A) with bulk and tapped densities of 0.25 ± 0.00 and 0.29 ± 0.00 g/mL, was able to absorb water and oil (17.31 ± 0.77% and 13.39 ± 0.29%). The powder was ICP-MS proved to be uncontaminated with heavy metals prohibited for cosmetics. The powder morphology was plate-like crystals with fractured as SEM observed (Figure 1B). Thermal properties examined by DSC resulted in glass transition, crystallization and melting temperatures of 66.43, 70.25 and 73.39°C, with 51.32 ± 4.13% of crystallinity by XRD. The color ($\Delta E \leq 1.43 \pm 0.04$), morphology, thermal properties, water absorption capacity and oil absorption capacity slightly shifted following to 3-month storage under ambient temperature and 45°C.

Introduction

Thanaka (*Hesperethusa crenulata* Roem.) has been being used in Myanmar and some parts of Thailand as the traditional cosmetics by grinding it on the flat stone moist with water and apply on the skin [1]. In the previous studies, there are several kinds of Thanaka research concerning anti-oxidant, anti-inflammatory, anti-microbial, skin lightening and sun protection properties [1-4]. Thanaka's specifications for cosmetic development are sparse to be presented, although its flowability was reported [1]. Of which, the bark was more suitable than the stem for solid dosage form development [1]. This research is to study Thanaka to be used as an absorbent alternative in the face powder since it contains tannins [5] which could have oil control property and it is not fully soluble in water [2]. Besides, Thanaka has a yellowish-white color which is suitable for Asian skin. However, other specifications of Thanaka for cosmetic development are meagerly informed. The objectives of this study aimed to analyze the physicochemical properties of Thanaka bark powder and examine its stability for applying in face powder products.

2. Experimental

2.1 Preparation of Thanaka Bark Powder

Thanaka stem was purchased from Mae Sai District, Chiang Rai. Thanaka bark was separated and was dried in a hot air oven at 45°C. It was finely ground and sieved to the particle size of powder < 45 μm.

2.2 Evaluation of Physicochemical Properties of Thanaka Bark Powder

2.2.1 Appearance, Particle Morphology and Color

Visual assessment was performed, and the particle morphology was examined by field emission gun-scanning electron microscopy [6]. The color was evaluated by UltraScan PRO spectrophotometer and recorded L^*, a^* and b^* [7].

2.2.2 Bulk and Tapped Densities & Cohesiveness and Flowability [8-9]

The Cohesiveness and the flowability were expressed triplicate in terms of the Hausner ratio and compressibility index respectively, where Hausner ratio is obtained as tapped density/bulk density, and the compressibility index is as the percentage of the density difference between tapped and bulk by tapped density. Bulk density was determined triplicate as the mass of Thanaka bark powder by apparent volume, m/V_0, in g/mL meanwhile, tapped density was continued measuring from bulk density as m/V_f, in g/mL.

2.2.3 Water and Oil Absorption Capacities [10-11].

Water and oil absorption capacities were measured slightly modified to the methods of Twinomuhwezi et al. (2020) and Chandra et al. (2015) by percent bound for 0.1 g of sample.

2.2.4 Thermal Properties and Crystallinity

The thermal properties were measured by Differential Scanning Calorimetry (DSC). The gelatinization onset (T_o), peak (T_m) and end (T_e) temperatures were analyzed [12]. The crystalline structure was examined by X-Ray Diffraction (XRD). Diffractograms were obtained from 5°-90° (2θ) at a scanning speed of 6°C/min [13]. The % crystallinity was calculated as $[100 \times [A_c/(A_c+A_a)]]$, where A_c is the area of crystalline peaks of diffraction, and A_a is the area of amorphous peaks of diffraction.

2.2.5 Elemental Analysis

The ICP-MSMS was used to detect As, Cd, Pb and Hg in Thanaka bark powder and calculated the result into parts per million [14].

2.3 Stability of Thanaka Bark Powder

The stability was measured at ambient temperature (TA) and 45°C for 3 months [1]. The appearance, particle morphology, color, water and oil absorption capacities, thermal properties and % crystallinity were evaluated by comparing with the initial.

3. Results and discussion

3.1 Evaluation of Physicochemical Properties of Thanaka Bark Powder

3.1.1 Appearance, Particle Morphology and Color

Thanaka bark powder is a fine and fluffy powder, and the color is yellowish light brown (Figure 1A) which L^*, a^* and b^* values, 75.63 ± 0.01, 5.58 ± 0.02 and 23.45 ± 0.02, respectively were corresponding. However, in the previous study the L^*, a^* and b^* of Thanaka bark powder from the different source were 81.49 ± 0.01, 2.21 ± 0.00 and 28.24 ± 0.11, which was slightly different [1]. The morphology was plate-like crystals with fracture as observed by SEM in Figure 1B.

3.1.2 Bulk and Tapped Densities & Cohesiveness and Flowability

Bulk and tapped densities were observed 0.25 ± 0.00 g/mL and 0.29 ± 0.00 g/mL, respectively. The bulk density was similar to the previous report, while that of tapped was lower [1]. The difference of tapped density shall be governed by the particle distribution as per surface and inherit moisture contents [8]. Flowability parameters in terms of compressibility index and Hausner ratio were calculated as 12.04 ± 0.24 and 1.14 ± 0.00, respectively, referring to less cohesiveness and good flowability of the powder in regard with the European Pharmacopoeia [8-9]. It should be noted that the flowability of this presenting study is different from the previous finding [1], at which particle distribution and moisture (surface and inheritance) delineated accordingly density and flowability.

3.1.3 Water and Oil Absorption Capacities

Thanaka bark powder absorbed water and oil efficiently (17.31 ± 0.77% and 13.39 ± 0.29%). These properties enhanced the consistency for adhering moisture from the surface/environment include sweat and sebum, that applicable for cosmetic solid dosage form [10-11, 15].

3.1.4 Thermal Properties and Crystallinity

The assessment revealed (Figure 1C) that the Thanaka bark powder started to be melted at 66.43°C (T_o), the melting ended at 73.39°C (T_e) and the gelatinization enthalpy was -0.28 J/g. At which the melting temperature of the powder was shown to be at 70.25°C (T_m). The low melting point of the powder shall be low in smoothness during topical application reduces spreadability accordingly, although its flowability is good. Nonetheless, the wearability will be good instead. Thus, those of solid dosage form should be more applicable for Thanaka bark powder than semi-solid one. That is in harmony with the crystallinity examined by XRD (51.32 ± 4.13%). Crystallinity refers texture of the material including thermal stability [16].

3.1.5 Elemental Analysis

In terms of contaminants, elemental analysis of Thanaka bark powder was undertaken to ensure the material's safety for topical product application. It should be noted that those of prohibited metals, i.e., Hg, Pb, As and Cd (0.00072 ± 0.00005, 0.03026 ± 0.00013, 0.21331 ± 0.01404 ppm and less than 0.00004 ppm, respectively) that are crucial for cosmetic products, were detected below the limitations as regulated by ASEAN guidelines [17].

3.2 Stability of Thanaka Bark Powder

After 3 months of stability assessment at ambient temperature (TA) and 45°C, Thanaka bark powder color was visually undetermined [7], although that of ΔE were 1.43 ± 0.04 and 1.26 ± 0.11, respectively [7]. In addition, the powder morphology was conformed with the initial state as exhibited in Figure 1B. Thus, densities and flowability parameters of powder should be stable in accordance with the exhibited morphology that governs the particle property of powder.

(A)

(B)

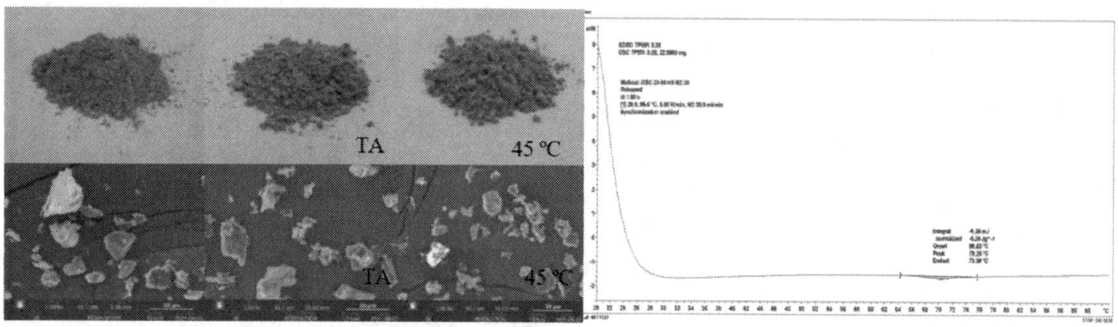

Initial After 3 months (C)

Figure 1. Thanaka bark powder **(A),** particle morphology by SEM **(B)** and thermal properties by DSC **(C)**.

The melting temperature of Thanaka bark powder became slightly higher temperature for both conditions. Furthermore, the water and oil absorption capacities and % crystallinity of Thanaka bark powder insignificantly ($p > 0.05$) changed after the stability test (Figure 2).

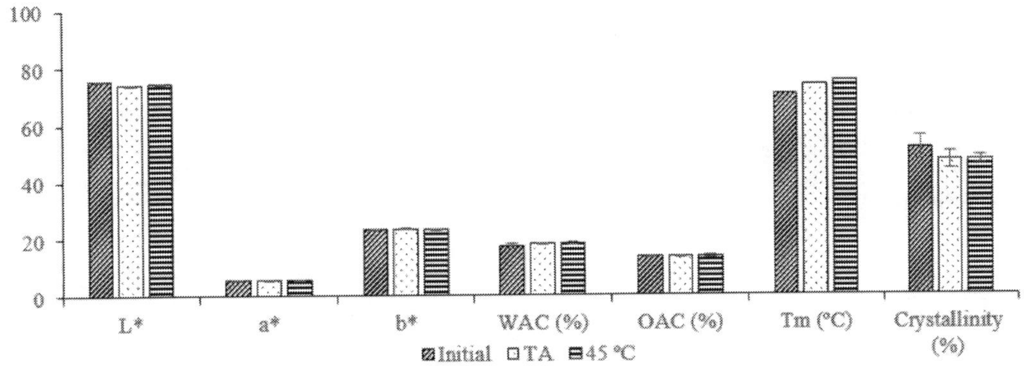

Figure 2. Stability test of color (*L**, *a** and *b**), water & oil absorption capacities (WAC & OAC), melting temperature (T_m, °C) and crystallinity (%) of Thanaka bark powder before and after 3 months at ambient temperature (TA) and 45°C.

4. Conclusions

Thanaka bark powder does not contaminate with restricted elements to a certain level. It has low cohesion and good flowability, and its water absorption and oil absorption efficacy have been confirmed. After the stability test, the color, morphology, water and oil absorption capacities, melting temperature and crystalline structure had some minor shifts. Therefore, this study results suggested that Thanaka bark powder may be potentially used as a natural raw material powder in color cosmetics, especially for face powder products.

Acknowledgment

This research study was financial and facilities supported by Mae Fah Luang University. The authors acknowledged MFU Student Scholarship 2020, Mae Fah Luang University upon partially financial support per study on the Master's Degree of Science Programme in Cosmetic Science.

References

[1] Kanlayavattanakul M and Lourith N 2012 *H&PC Today*. **2** 30-32.
[2] Wangthong S, Palaga T, Rengpipat S, Wanichwecharungruang S P, Chanchaisak P and Heinrich M 2010 *J. Ethnopharmacol*. **132** 466-472.
[3] Jantarat C, Sirathanarun P, Chuchue T, Konpian A, Sukkua G and Wongprasert P 2018 Sci *Pharm*. **86** 1-9.
[4] Joo S H, Lee S C and Kim S K 2004 *J. Plant Biol*. **47** 163-165.
[5] Aung N N 2006 Production of Traditional Myanmar Thanakha as cosmetic, *University of Yangon*, 64 pages.
[6] Almeida C S, Barretti R V, Ito V C, Malucelli L, Filho M A S C, Demiate I M, Pinheiro L A and Lacerda L G 2020 *Mater. Res*. **23** 1-9.
[7] Lourith N and Kanlayavattanakul M 2012 *Adv. Sci. Lett*. **5** 1-4.
[8] European Pharmacopoeia 6.0 2008 Powder Flow, *Strasbourg: Council of Europe*, 4 pages.
[9] European Pharmacopoeia 7.0 2010 Bulk Density and Tapped Density of Powders and Measures of Powder Compressibility, *Strasbourg: Council of Europe*, 4 pages.
[10] Twinomuhwezi H, Awuchi C, Rachael M 2020 *AJFSN*. **1** 6-19.
[11] Chandra S, Singh S and Kumari D 2015 *J. Food Sci. Technol*. **52** 3681-3688.
[12] Morales-Scanchez E, Figueroa J D C and Gaytan-Martinez M 2009 *J. Food Sci*. **74** 382-385.
[13] Jaiturong P, Laosirisathian N, Sirithunyalug B, Eitssayeam S, Sirilun S, Chaiyana W and Sirirthunyalug J 2020 *Heliyon*. **6** 1-9.
[14] Tanase G, Udristoiu F M, Bunaciu A A and Aboul-enein H Y 2012 *Gazi University Journal of Science*. **25** 843-851.

[15] Abdullah E C, Salam A M and Aziz A R 2010 *Phy. Intl.* **1** 16-21.
[16] Ashton Acton Q 2013 (Eds.) Eye Proteins-Advances in Research and Application, *ScholarlyEditions*, 376 pages.
[17] https://asean.org/wp-content/uploads/2012/05/ASEAN-Guidelines-Limits-of-Contaminant-Cosmetics-.pdf; May 2017.

Utilization of lignin extracted from Thai ago-waste as UV-blocking agent for BG-lignin/PLA composite films

P Wijaranakul[1]*, B Hararak[2]*, C Winotapun[2], P Wannid[2], C Chotirotsukon[3] and W Laosiripojana[1]

[1] Department of Tool and Materials Engineering, Faculty of Engineering, King Mongkut's University of Technology Thonburi, Bangkok 10140, Thailand
[2] National Metal and Materials Technology Center (MTEC), National Science and Technology Development Agency, Pathumthani 12120, Thailand
[3] National Center for Genetic Engineering and Biotechnology (BIOTEC), National Science and Technology Development Agency, Pathumthani 12120, Thailand

Corresponding author: pawarisa.wijaranakul@mail.kmutt.ac.th and bongkoth@mtec.or.th

Abstract. The improving properties of poly(lactic acid), PLA, by utilizing natural resources attracted great intrigue to make such a green composite material that can be used as a commercial product in human life [1]. Lignin is one of biopolymer that can be used as bio-based filler and multifunctional bio-additive in a polymer composite. The most exciting properties of lignin that can be provided in a polymer composite is UV absorption and anti-oxidation [2]. In this current study, organosolv lignin extracted from sugarcane bagasse (BG-lignin) was utilized as multifunctional bio-additive in PLA for improving the UV absorption. The physicochemical and thermal properties of BG-lignin were determined using several techniques including SEM, GPC, quantitative ^{31}P NMR, and DSC. BG-lignin at different loading contents (0.1, 0.2, 0.5, and 1 wt.%) was mixed with PLA via melt-extrusion. The attained compounds were converted to composite films via blown film extrusion. With the 0.5 wt.% loading content, the PLA composite films (0.5BG-lignin/PLA) absorbed almost all UV radiation which exhibits almost 70% blocking of UVB. The onset oxidation temperature of the PLA/0.5BG composite film increased by 34% as compared to that of the neat PLA film. Adding of BG-lignin enhanced tensile strength and Young's modulus but did not favour to elongation at break.

1. Introduction

Ultraviolet (UV) radiation is high-energy-visible light with a wavelength from 200 to 400 nm and present in sunlight about 10% of total electromagnetic radiation; the long term of UV radiation exposure will cause photodegradation of polymeric materials [3]. Therefore, being an improvement of UV-shielding polymer composite is a desirable interest. Lignin is a polyphenolic compound and the second most abundant renewable biomaterial after cellulose. It is the main component of the plant cell wall to the cellulose and hemicellulose for mechanical support in their structure [4]. Lignin has an aromatic structure and many functional groups such as hydroxyls, methoxy, carbonyl, and carboxyl, etc. In the academic area, lignin has been recognized as the bio-multifunctional additive that can develop plastic composite properties, especially, antioxidant [2] and UV-shielding ability [5]. Sugarcane bagasse is one of the most agricultural waste in Thailand. It could also be one of the most promising resource for lignin extraction. Utilization of lignin for high value-added applications is one of the great challenges, especially in the application for a biodegradable single-use plastics. Poly(lactic acid), PLA, has become an alternative choice to replacing petroleum-based plastics. However, poor UV absorption of PLA films might not be suitable for UV-sensitive products. Using of lignin as UV-absorber in PLA films sounds to be a promising choice to attain fully biodegradable PLA composite films.

The objective of this current work is to utilization of organosolv lignin extracted from sugarcane bagasse as multifunctional bio-additives for PLA films. BG-lignin/PLA composite films were prepared via ubiquitous melt-extrusion. Different contents of BG-lignin (0.1, 0.2, 0.5, and 1 wt.%) were incorporated into PLA. Properties of the obtained BG-lignin/PLA composite films including film color, mechanical properties, UV-blocking ability, and antioxidant activity were investigated in comparison with the neat PLA film.

2. Experimental

2.1. Materials

Organosolv sugarcane bagasse-lignin (BG-lignin) was kindly supported from the Biorefinery and Biological Technology Research Group (National Center for Genetic Engineering and Biotechnology (BIOTEC)). PLA (4043D) used in this study is a product of NatureWork LLC. Its number-average molecular weight (M_n) is 160,000 g/mol containing 94% L-lactic acid content and the melt flow rate is 6.0 g/10 minutes at 210°C.

2.2. Characterizations of BG-lignin

The morphology of BG-lignin was observed by field-emission scanning electron microscope (FE-SEM, model SU5000, Hitachi, Japan). The molecular weight distribution was thoroughly analyzed by gel permeation chromatography (GPC, Waters e2695 separation module, USA). Quantitative phosphorus-31 nuclear magnetic resonance spectroscopy, ^{31}P NMR, (AV-500 Bruker Biospin, USA) were used for determining chemical characteristic. Protocol of ^{31}P NMR solution preparation was followed steps reported in Nature protocol [7]. Glass transition temperature (T_g) of the BG-lignin was measured in a temperature scan at heating rate of 10°C/minute using differential scanning calorimetry (DSC-1, Mettler Toledo, Switzerland).

2.3. Preparation of BG-lignin/PLA composite films

Pre-dried PLA pellets were premixed with the 2 wt.% of BG-lignin powders prior to load to a feeder of the single-screw extruder (Thermo-Haake Rheomix OS, Germany) equipped with 3-mm rod die. Temperature profile was set as 170, 180, 185°C and die temperature was 180°C. Rotating speed was 45 rpm. The extrudates were extruded through a water bath and was then pelletized into 3-mm pellets. Attained 2 wt.% BG-lignin/PLA pellets were diluted to the desired loading contents (0.1, 0.2, 0.5, and 1 wt.%) by adding of pristine PLA. BG-lignin/PLA composite films were manufactured using the same single-screw extruder equipped with blown film (34-mm inner diameter and 1-mm die gap). Screw speed was kept at 60 rpm to fabricate film with the width and thickness of 13 cm and 35 μm, respectively.

2.4. Characterizations of BG-lignin/PLA composite films

Properties of the obtained neat PLA film and its composite films with BG-lignin/PLA were conducted. Color of the films was determined by datacolor spectrophotometer (datacolor 650, Pakisatan). Mechanical properties including tensile strength, Young's modulus and elongation at break were performed using a universal testing machine (Instron 4502 series, USA) by following ASTM D882 at the cross-head speed of 400 mm/10 minute. Light transmittance of all the films was measured using a UV spectrometer (UV-3600Plus, Japan) ranges from 200-800 nm. The oxidative induction temperature (OIT_{temp}) was measured by using DSC under oxygen atmosphere, a DSC protocol as described in literature [8].

3. Results and discussion

SEM images of BG-lignin are presented in Figure 1. BG-lignin particles seem to be constructed by the accumulation of small lignin particles (see Figure 1a). The surface and shape of BG-lignin particles were evaluated at high magnification of 20,000x (see Figure 1b). BG-lignin were irregular in their shape, which varied diameter from sub-micrometer to a few microns. Molecular weight (M_w) and polydispersity index (PDI) of BG-lignin were evaluated by the GPC technique using the polystyrene as a standard polymer. M_w and the PDI of BG-lignin are 1,638 g/mol and 1.62, respectively. The phenolic hydroxy group was considered as an essential functional group due to the UV-absorption of its active group, which was investigated by ^{31}P NMR, BG-lignin exhibits high content of phenolic to aliphatic hydroxyl groups, which is 1.82 mmol/g. The content of the phenolic hydroxyl group could be correlated to the active functions of lignin including UV-shielding and antioxidant [2]. T_g of BG-lignin was detected at 135°C.

Figure 1. SEM images of the BG-lignin particles at (a) 2,000x and (b) 20,000x.

Clear neat PLA film and brownish BG-lignin/PLA composite films were observed. Chromatic coordinate was employed to determine the effect of BG-lignin on PLA film color and transparency. The neat PLA film exhibited L*, a*, b* at 90.41, -0.52, and 11.51, respectively. All these values were used as the reference for testing on the BG-lignin/PLA composite films containing different contents of BG-lignin. The neat PLA film exhibited the highest L* value, while that of the BG-lignin/PLA composite films decreased with an increase of the BG-lignin contents. Color change represented in ΔL, Δa, Δb was calculated based on the neat PLA film. At the highest loading contents of 1.0 wt.% (1.0BG-lignin/PLA), the values of ΔL, Δa, Δb became -4.69, 1.48 and 7.50, respectively. This could be implied that the composite film became darker and the color turned to brownish

Mechanical properties of the neat PLA film and its BG-lignin composite film were measured and compile in Table 1. Tensile strength and Young's modulus of the BG-lignin/PLA composite films increased as compared to the values of the neat PLA film. This increasing trend reveal to the high stiffness aromatic structure of lignin, and it possible due to the good interfacial adhesion between BG-lignin and PLA matrix [5]. The highest tensile strength at 65.0±2.0 MPa was found when the BG-lignin was 0.1 wt.%. The highest Young's modulus at 2.0±0.3 GPa was detected on the composite film containing 0.2 wt.% BG-lignin. Further increasing of BG-lignin, tensile strength and Young's modulus were decreased.

Table 1 Tensile properties and anti-oxidation ability of neat PLA and its BG-lignin composite films.

Samples	Tensile strength at break (MPa)	Young's modulus (GPa)	Elongation at break (%)	OIT_{temp} (°C)	Anti-oxidation ability (%)
Neat PLA	54.5±4.5	1.3±0.4	5.0±0.3	215	0
0.1BG-lignin/ PLA	65.0±2.0	1.7±0.1	4.8±0.6	247	15
0.2BG-lignin/ PLA	63.9±2.0	2.0±0.3	4.4±0.1	271	25
0.5BG-lignin/ PLA	62.7±2.2	1.7±0.1	4.5±0.2	289	34
1.0BG-lignin/ PLA	56.6±1.4	1.5±0.3	5.0±0.6	299	38

UV-visible spectra of the neat PLA and BG-lignin/PLA composite films were plotted as a function of wavelength as shown in Figure 2. The neat PLA film exhibits a nearly transparent character at 90% transmittance in visible region (400-800 nm). High 90% transmittance was still detected in UV region, this indicates poor UV absorption of PLA film. Adding small amount of BG-lignin, the composite films show an improvement of UV absorption. It is due to the phenolic hydroxyl groups of the BG-lignin. Increasing of BG-lignin effected directly to increase the UV absorption, while the transparency was satisfied [2]. However, loading content of BG-lignin resulted in film color as more brownish was noticed. Composite film with 0.5 wt.% BG-lignin (0.5BG-lignin/PLA) is expected as a suitable film which exhibits almost 70% blocking of UVB (315-280 nm).

As the results shown in Table 1, antioxidant activity for the neat PLA and its composite films was considered using OIT_{temp}. The OIT_{temp} of each film was evaluated on the onset temperature of exothermal characteristic. An antioxidant activity (%) was calculated based on the neat PLA film. Additional BG-lignin can enhance significantly on the OIT_{temp} and the anti-oxidation activity. This can be explained by the phenolic hydroxy groups in lignin acting as free radical scavenger that can promote anti-oxidation activity of the composite films.

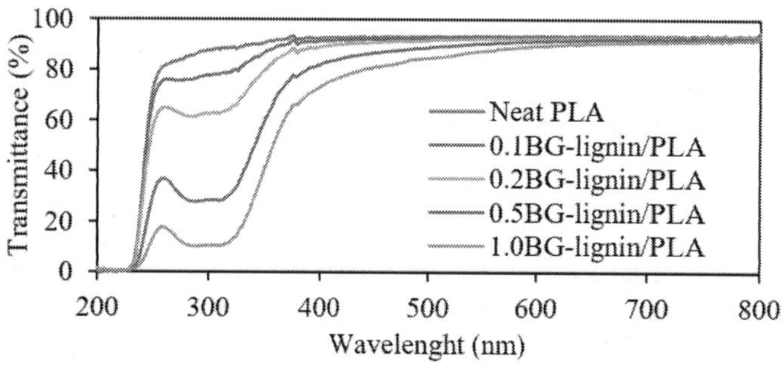

Figure 2. UV-Vis spectra of neat PLA and its BG-lignin/PLA composite films.

4. Conclusions

In this current work, we have demonstrated the utilization of lignin extracted from Thai ago-waste, sugarcane bagasse, as multifunctional bio-additive in PLA film. BG-lignin/PLA composite films achieve excellent in UV-shielding ability and anti-oxidation. The composite film with 0.5 wt.% BG-lignin (0.5BG-lignin/PLA) exhibits almost 70% blocking of UVB while film transparency and color are satisfied. The OIT_{temp} of the 0.5BG-lignin/PLA film become to 289°C which is 74°C higher than that of the neat PLA. Mechanical properties of the composite films were kept in the acceptable criteria. Therefore, BG-lignin can be an alternative bio-additive providing multifunctional for PLA films.

Acknowledgement
Authors appreciate the funding support from MTEC, NSTDA, Thailand (project no.2052415). The granted from Program Management Unit Competitiveness (PMUC) "Lignin valorization for production of biochemicals and biomaterials in biorefinery".

References
[1] Ncube L K, Ude A U, Ogumuyiwa E N, Zulkifli R. and Beas I. N. 2020 materials **13** 4994.
[2] Cavallo E, He X, Luzi F, Dominici F, Cerrutti P, Bernal C, Foresti M L, Torre L and Pugli D 2020 molecules **26** 126.
[3] Lu T, Solis-Ramos E, Yi Y and Kumosa M 2018 Polymer Degradation and Stability **154** pp. 203-210.
[4] Schneider W D H, Dillon A J P and Camassola M, 2021 Biotechnology Advances **47** 107685.
[5] Hararak B, Winotapun C, Inyai J, Wannid P and Prahsarn C 2021 J. Nanopart Res **23** 193.
[6] Meng X, Crestini C, Ben H, Hao N, Pu Y, Ragauskas A J andArgyropoulos D S 2019 natural protocols **14** pp 2627-2647.
[7] Sadeghifar H and Argyropoulos D S 2015 ACS Sustainable **3** pp 349-356.

Effect of arecoline, aqueous and methanolic areca nut extracts on rumen fluke

N Puyathorn[1], C Sukonpan[1], P Toungsuwan[2], W Narakornwit[1], T Phaechamud[1*]

[1]Faculty of Pharmacy, [2]Faculty of Education, Silpakorn University, Nakhon-Pathom, Thailand. 73000.

*Corresponding author: phaechamud_t@su.ac.th

Abstract. Paramphistomosis is a parasitic disease that causes a reduction of livestock production and economic losses worldwide. In order to avoid side effect and chemical residue from usage of chemical drugs, medicinal plant is an alternative drug for treatment. Areaca nut (*Areaca catechu*) which one of the domestic plants in Southeast Asia had antihelmintic activity. In this study, arecoline, aqueous and methanolic extracts of areca nut prepared in solutions were tested for antihelmintic activity against rumen flukes *(Paramphistomum sp.)*. The parasites were incubated in a culture media (RPMI-1640, pH 8) at 39°C in a carbon dioxide-incubator containing 0.01, 0.1 and 1 mg/mL of the tested compounds for 15, 30 min, 1 h and 24 h. The results showed that the methanolic and aqueous crude extracts from seeds of *Areaca catechu* and arecoline HBr at concentration of 1 mg/mL solutions could destroy the tegument (surface membrane) of the flukes and killed them all at initial time. Thus, these areca nut extracts exhibited the potential alternative medicine with antihelmintic activity against rumen flukes.

1. Introduction

Paramphistomosis is an infectious disease caused by rumen fluke in *Paramphistomum sp.* This disease is one of a serious disease occurred in buffaloes, cattle, goats and sheep worldwide [1]. This parasitosis has been found in domestic and wild ruminants caused by trematode. In the early stage of infection, young flukes are located in the small intestine and abomasums; subsequently they move to the rumen to finally lodge as adult trematodes [2,3]. The parasites caused the infected animals to diminish their appetite and delay the growth from malnutrition and death finally if untreated. This leads to a reduced livestock production and an economic loss [4]. Excessive usage of chemical drug will increase a production cost and leave some chemical residues in the animal products. To avoid these effects, herbal medicine is interested [5]. *Areca catechu* Linn or areca palm is an endemic plant in tropical Pacific and Southeast Asia. Its seeds that called areaca nut or betel nut or "MAK" in Thai (Figure 1A and B) had been chewed along with betel leaf and also used in Ayurvedic medicine since ancient times [6]. One of major active alkaloids found in areaca nut is arecoline (methyl-1, 2, 5, 6-tetrahydro-1-methyl-nicotinate) (Figure 1C) which has various pharmacological activities [7]. One of the interesting activities is antihelmintic activity. Arecoline HBr is a white crystalline powder or needle-shaped crystal with a melting point of 169-171°C, odorless and bitter in taste, soluble in water and alcohol [8]. Arecoline (Figure 1C) is used as veterinary medicine as antihelmintic compound to diminish parasites [7,8]. It was found that betel nut extract was able to control the liver fluke of goats. The aim of this study was to investigate the effect of aqueous and methanolic crude extracts from areca nut as well as arecoline HBr on mobility and modification of tegument of rumen flukes.

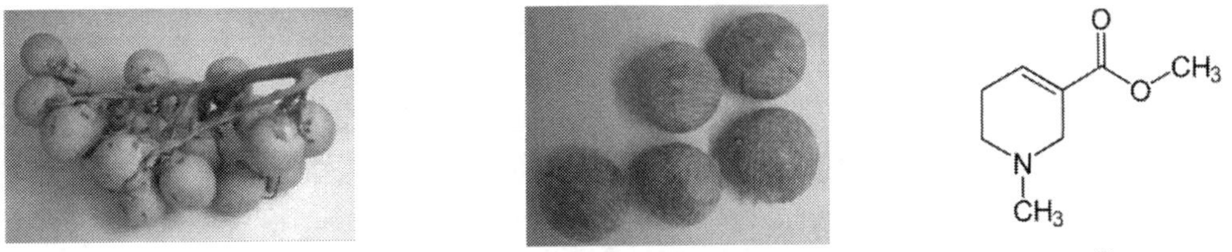

A B C

Figure 1. Ripe areca nuts (A); areca seeds (meat part) after removed all fibrous husk (B) and chemical structure of arecoline, a major active alkaloid found in areaca nut (C).

2. Experimental

2.1 Collection and selection of rumen fluke

The adult rumen flukes were selected and treated before use as following procedure. Live rumen flukes (*Paramphistomum sp.*) were directly collected from the rumens of infected cattle from a local/domestic slaughter house in Nakhon Pathom, Thailand. The collected fluked were cleaned with 0.85 % NaCl solution for several times prior to immerse in the culture media (RPMI-1640, pH 8) and subsequently incubated at 39°C in a carbon dioxide-incubator (CB150/210, BINDER, England).

2.2 Preparation of crude extracts of areca seed

Fresh ripe areca nuts from local orchard in Nakhon Pathom, Thailand were collected and its voucher specimen has been deposited at the Department of Pharmacognosy, Silpakorn University, Nakhon Pathom. The nuts were cleaned and removed all fibrous husk to get the seed. The nut and meat part inside the husk are shown in Figure 1A, B. The seeds were chopped into small pieces and dried in a hot air-blowing oven at 50°C. The dried samples were subsequently powdered by passing through 30 mesh stainless screen of cutting mill and kept at 4°C. Methanol extract and water extract of areca seed were prepared by immersing the seed powder in methanol and distilled water, respectively, in the ratio of the powder to solvent 1:2 for 24 h. The solvents were removed under the vacuum until dry. The crude extracts were storage in dry and cool place prior to test. Arecoline HBr (Figure 1C) was purchased from Sigma chemical, USA.

2.3 Antihelmintic activity of crude extracts of areca seed and arecoline

Antihelmintic activity was tested by two studies: movement behavior and modification of tegumental surface after exposing to the extracts at different designed times. About 50 adult flukes were placed into 10 petri dishes. One plate was assigned as a control group that did not added any sample solutions and the rest were experimental group that separately treated the flukes with the aqueous and methanolic extracts of areca seed and arecoline HBr at the concentrations of 0, 0.01, 0.1, 1.0, and 10 mg/mL, respectively. Movement behavior and physical appearance of the flukes in each plate were visually observed and recorded at 15, 30, 60 min, 12 h and 24 h, respectively. After finished the final observation, the flukes from the control group and the experiment groups were taken and soaked with Bouin's solution for fixing prior to further preparing paraffin embedded specimens. At the end of the embedded process, the flukes in the paraffin block were cut by rotary microtome to get 6 µm-thick specimens. After paraffin removing and rehydration, the specimen sections were submitted to Harris's hematoxylin staining and further observed tegumental modification of the flukes under a stereoscope (Miotic SMZ-171 Series, Hong Kong).

3. Results and discussion

The percentage yields of aqueous and methanolic crude extracts of areca nut were 23.35% and 23.41%, respectively. After incubated the flukes in media containing the extracts and arecoline HBr with various concentrations, the changing of movement behavior and physical appearances of the flukes were observed at assigned time (as shown in Table 1). At 1 mg/mL concentration, two crude extracts and arecoline HBr could kill all flukes in the plates. All parasites stopped moving immediately after adding the test solutions and their bodies were swelling after left them in the extract solutions for 24 h (Figure 2 C). At 0.1 mg/mL concentration, they showed a slightly different result after 24 h of incubation. All parasites stopped moving immediately after adding the test solutions but, at 24 h-incubation, about 10.7% and 14.5% of parasites in the plates containing aqueous extract and methanol extract, respectively, started moving slowly. At 0.01 mg/mL concentration, arecoline HBr showed better result in stopping parasite mobility. Arecoline HBr could halt the moving for longer than the extracts did since this substance has been reported as antihelmintic compound [7,8]. Only 5% of parasites recovered and started moving slowly after incubated them with 0.01 mg/mL of arecoline for 24 h while all of parasites incubated with the extracts started moving again at 15 min (Table 1). The results from the regressive hematoxylin staining cross-section of the fluke demonstrated that the parasites were killed from the tegument destruction [9,10]. Comparing to a smooth and thick tegument (tg) in control group in Figure 3, damage tegument's surface as in Figure 4 was observed in the fluke treated with 1 mg/mL aqueous areca nut extract. According to the chemical composition study by gas chromatography/mass spectrometry (GC-MS), areca nut had volatile methyl esters (arecoline and guvacoline) and six chemically related alkaloids (methyl nicotinate, ethyl nicotinate, nicotine, davidine and annine) [7]. The chemicals in aqueous and methanolic areca nut extracts might contain these compounds especially arecoline and influence on rumen fluke mobility with tegumental modification. The insecticide and lavicidal activities of *Areca catechu* seed have been reported previously [11,12]. The high tannin content of its aqueous and methanolic extracts as 42.41% and 47.72% with total phenolic of 42.06 ± 2.00 and 45.12 ± 2.04 g/100g crude extract as gallic acid, respectively [12], could be related with tegument modification of rumen fluke from its ability for protein precipitation.

Table 1 Effects of aqueous and methanolic crude extracts and arecoline of areca nut and arecoline on physical appearance and movement behaviour of rumen fluke.

Tested compounds	Physical appearance and movement behavior of rumen fluke								
	Aqueous extract			Methanolic extract			Arecoline		
Time	1 mg/mL	0.1 mg/mL	0.01 mg/mL	1 mg/mL	0.1 mg/mL	0.01 mg/mL	1 mg/mL	0.1 mg/mL	0.01 mg/mL
15 min	No move (100%)	No move (100%)	Shrink but still move (100%)	No move (100%)	No move (100%)	Shrink but still move (100%)	No move (100%)	No move (100%)	No move (100%)
30 min	No move (100%)	No move (100%)	Move (100%)	No move (100%)	No move (100%)	Move (100%)	No move (100%)	No move (100%)	No move (100%)
1 h	No move (100%)	No move (100%)	Move (100%)	No move (100%)	No move (100%)	Move (100%)	No move (100%)	No move (100%)	No move (100%)
24 h	No move and swell (100%)	Slowly move (10.07%)	Move (100%)	No move and swell (100%)	Slowly move (14.5%)	Move (100%)	No move (100%)	No move (100%)	Slowly move (5.0%)

*data in the parentheses are the percentage of worms with the above appearance

4. Conclusions

Methanolic and aqueous crude extracts from seeds of *Areaca catechu* could destroy the tegument of the flukes and killed them; therefore, these areca nut extracts showed the potential antihelmintic activity against rumen flukes. The arecoline and phenolic content especially tannin in these extracts should be active compounds promoting tegument modification of rumen flukes. Nevertheless, their adverse reactions and side effects on cattle should be monitored and considered.

IOP Conf. Series: Materials Science and Engineering 1234 (2022) 012019 doi:10.1088/1757-899X/1234/1/012019

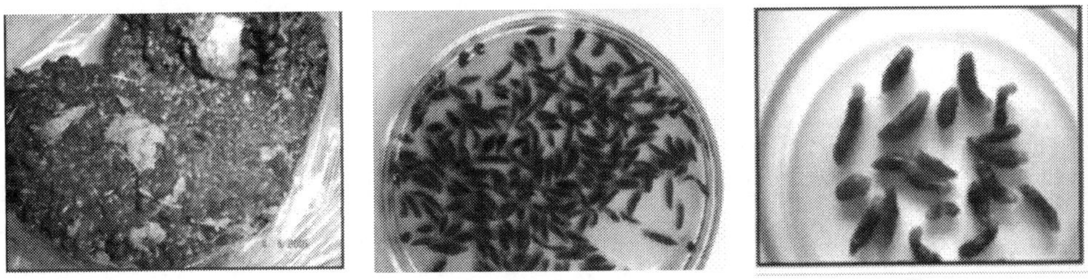

A B C

Figure 2. Physical appearance and fluke mobility (A. Mixed rumen flukes in cattle gut; B. Rumen flukes in control group had pear-shape and moved around.; C. Rumen flukes treated with 1 mg/mL water extract had swollen body and stopped moving).

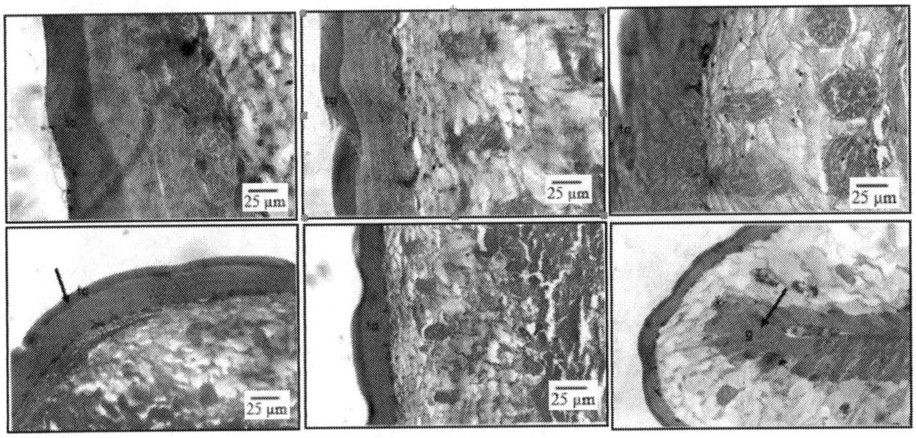

Figure 3. Smooth and thick tegument (tg) and gut (g) in a rumen fluke of control group.

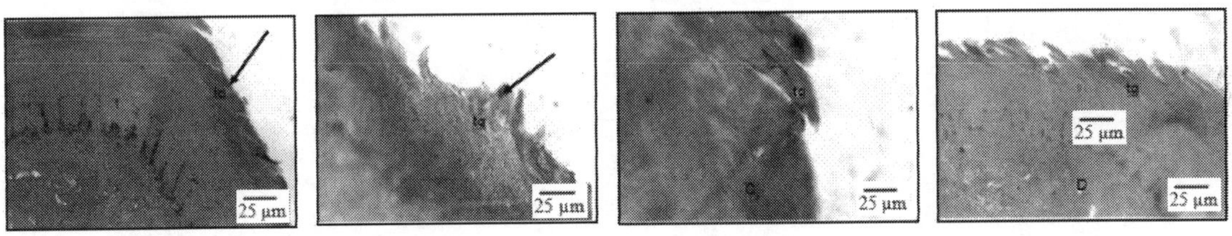

Figure 4. Destroyed tegument (tg) found in rumen fluke treated with 1 mg/mL aqueous areca nut extract.

Acknowledgements
The research team is grateful for the Research and Creative Fund, Faculty of Pharmacy, and Faculty of Education, Silpakorn University. The authors would like to thank Prof. Dr. Duangduen Krailas and Suluck Namchote for your support and help.

References
[1] Sanabria REF and Romero JR 2008 *Helminthologia* ***45*** 64 - 68.
[2] Rangel-Ruiz LJ, Albores-Brahms *ST* and Gamboa-Aguilar J 2003 *Vet. Parasitol.* ***116*** 217–222.
[3] Yamaguti S 1971 *Digenea of Mammals Synopsis of digenetic trematodes of vertebrates* Vol. I, Keigaku Publishing Company, Tokyo. Japan. pp. 695-711.
[4] Schell SC 1970 , *How to Know the Trematodes: Pictured Key Nature Series* W.C. Brown Company, Dubuque Iow., USA. pp. 1-331.
[5] Szmidt V, Abrous M, Adjidé CC and Dreyfuss G 2000 *Vet. Parasitol.* ***87*** 133–138.
[6] Oxenham MF 2002 *J. Archacol. Sci.* **29** 909-915.
[7] Holdsworth DK, Jones RA, and Self R 1998 *Phytochem.* ***48*** 581-582.

[8] Mui NT, Binh DV and Orskov ER 2005 *Animal Feed Sci. Technol.* ***121*** 77-87.
[9] Coustau C and Yoshino TP 2000 *Exp. Parasitol.* ***94*** 62-66.
[10] Horak IG 1971 *Adv. Parasitol.* **9** 33-72.
[11] Chu NS 2001 *J. Biomed. Sci.* ***8*** 229-236.
[12] Wetwitayaklug P, Phaechamud T, Limmatvapirat C and Keokitichai S 2006 *Naresuan University J.* ***14*** 1-14.

The validation HPLC method for determination of gambogic acid in gamboge resin

E M Khaing[1], K Jitrangsri[2], J Mahadlek[3], C Sukonpan[4] and T Phaechamud[1,5*]

[1]Program of Pharmaceutical Engineering, Faculty of Pharmacy, Silpakorn University, Nakhon Pathom, Thailand
[2]Program of Pharmaceutical Sciences, Faculty of Pharmacy, Silpakorn University, Nakhon Pathom, Thailand
[3]Pharmaceutical Intellectual Unit "Prachote Plengwittaya", Faculty of Pharmacy, Silpakorn University, Nakhon Pathom, Thailand
[4]Department of Pharmaceutical Chemistry, Faculty of Pharmacy, Silpakorn University, Nakhon Pathom, Thailand
[5]Natural Bioactive and Material for Health Promotion and Drug Delivery System Group (NBM), Faculty of Pharmacy, Silpakorn University, Nakhon Pathom, Thailand

*Corresponding author: phaechamud_t@su.ac.th

Abstract. Gambogic acid obtained from plant resin of *Garcinia* species has anticancer properties. This study focuses on the development and validation of the high-performance liquid chromatography (HPLC) method for the quantification of gambogic acid (GA) content in the gamboge raw sample. The chromatographic system was performed on C18 column (150 × 4.6 mm, 5 μm). The system consisted of acetonitrile and 0.1% orthophosphoric acid (85:15 v/v) as mobile phase with a flow rate of 1.5 mL/min. The detection was carried out at a wavelength of 360 nm and the retention time of GA was evident at 9 min. The method was validated according to ICH guidelines. The obtained calibration plot exhibited good linearity in the range of 5 to120 μg/mL ($r^2 > 0.999$). The percentage recovery GA spiked in the gamboge sample ranged from 98.87% to 102.92%. The LOD and LOQ were shown to be 2.069 and 6.271 μg/mL, respectively. The validation study demonstrated that this method was simple, specific, and applicable for quantitative analysis of GA in gamboge resin.

1. Introduction

Gamboge, plant resin is obtained from the *Garcinia* species such as *Garcinia hanburyi* Hook.f., has been utilized in colour pigmentation and folk medicine for the treatment of pain, edema, infected wound and laxative [1]. Gambogic acid (GA) is the caged xanthone family and is isolated from gamboge resin as the main active compound. Its antitumor activities are interested in pharmaceutical applications. GA has a potent anticancer effect against various cancer cells containing liver, breast, lung, colon, prostate and ovarian cancer [2]. The mechanism of GA on colon cancer SW620 cells was due to the suppressing of cell proliferation and dispersion in a dose-dependent effect *in vitro* assay [3].

HPLC method was explored for the determination of eight polyprenylated xanthone compounds in gamboge resin including gambogic acid [4]. The determination of GA and gambogenic acid in rat plasma could be conducted using liquid chromatography–tandem mass spectrometry [5]. A non-aqueous capillary electrophoresis method has been developed to determine gambogic acid in gamboge [6]. However, only GA content analysis in gamboge resin using HPLC method has not been reported in the literature. Thus, this study aimed to develop a simple and specific HPLC method for the quality control of GA content in gamboge resin and further apply it for the assessment of GA content in pharmaceutical formulations.

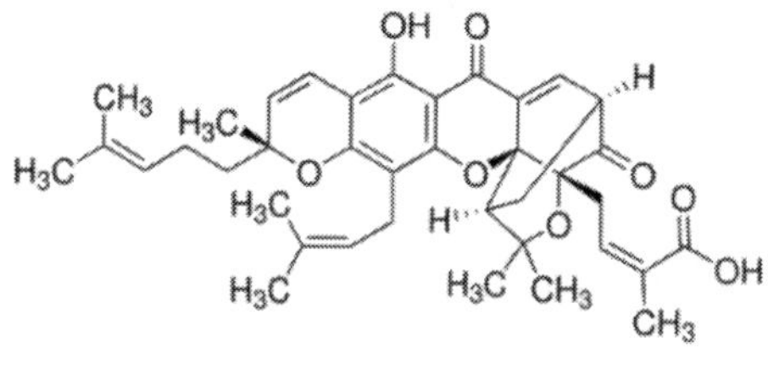

Figure 1. Structure of gambogic acid.

2. Experimental

2.1. Materials

Gambogic acid, purity >98% (Chengdu Biopurify Phytochemicals Ltd., China) was employed as received and gamboge resin was obtained from a local drug store, Thailand. HPLC grade Acetonitrile, methanol (RCI Labscan) and orthophosphoric acid (Ajax Finechem) were purchased from CT chemical Ltd., Bangkok, Thailand.

2.2. Instrumentation and chromatographic condition

Agilent 1260 HPLC system with a photodiode array detector (Infinity, USA) and C_{18} column (150 × 4.6 mm, 5 μm particle size, Dr. Maisch, Germany) were used. The mobile phase composed of 85% acetonitrile and 15% orthophosphoric acid (0.1%). The mobile phase was filtered through 0.45 μm membrane filter before use and sonicated for 20 min. The isocratic elution was performed at 1.5 mL/min flow rate with a 12 min of run time. The column temperature was processed at ambient temperature. The sample injection volume was 20 μl and the detection wavelength was at 360 nm.

2.3. Preparation of standard solutions

The accurately weighed amount of GA (10 mg) was dissolved in methanol at a 1 mg/mL concentration using a volumetric flask. The solution was sonicated for 10 min and adjusted to the volume with methanol. Then, the working standard solution was prepared with phosphate buffer solution pH 7.4, to obtain the final concentration of 5 - 120 μg/mL.

2.4. Method Validation

The validation of HPLC method was performed in accordance with ICH guidelines for specificity, linearity, accuracy, limit of detection (LOD), limit of quantification (LOQ) and precision (intra-day and inter-day variation) [7].

2.5. Preparation of GB sample

The gamboge sample was accurately weighed and dissolved in methanol and mixed well before filtration through the membrane filter (0.45 mm). The solution was further diluted to the appropriate concentration. Three different gamboge samples were used for the determination of GA content.

3. Results and discussion

3.1. Method Validation

Specificity

The chromatogram of GA standard solution is presented in Figure 2a. The retention time of the method was approximately 9 min. This developed method provided a shorter analysis time than the previous method [4].

Linearity
The method demonstrated good linearity with a correlation coefficient (r^2) of 0.9998. The slope and intercept were calculated using linear regression analysis. The equation was y = 11.803x + 3.1421. The linear relationship was evident between the concentration of analyte and the peak area (Figure 3). The linearity range for GA was 5-120 µg/mL.

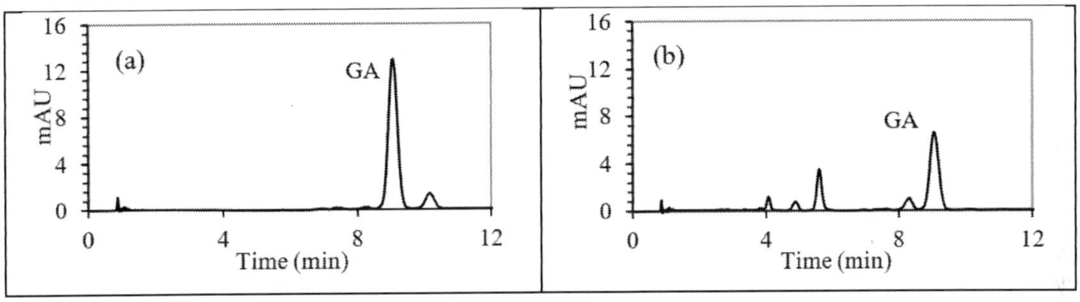

Figure 2. HPLC chromatograms of GA in standard solution (25 µg/mL) (a) and gamboge resin (b).

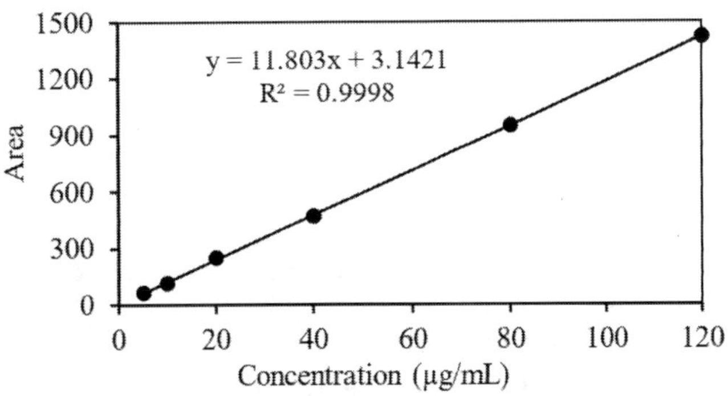

Figure 3. Linearity of standard gambogic acid detected at the wavelength 360 nm.

Accuracy
The accuracy was investigated as a standard addition method. The known concentration of gamboge sample was prepared by spiking with 8, 10 and 12 µg/mL of gambogic acid solution. The mixture solutions were analyzed and calculated for recovery (%) and relative standard deviation (% RSD) for each concentration. The validated method was shown to be accurate and % RSD of each level was less than 2%. The results are summarized in the Table 1.

Table 1. Recovery data of GA at concentration of 8, 10, 12 µg/mL (n = 3).

Concentration (µg/mL)			% Recovery (Mean ± SD)	% RSD
Spiked	Sample	Calculated spiked (Mean ± SD)		
8	10	8.100 ± 0.077	101.1 ± 0.96	0.95
10	10	9.828 ± 0.095	98.87 ± 0.95	0.97
12	10	12.344 ± 0.068	102.92 ± 0.57	0.55

Precision

The intra-day precision was obtained by determination of six replicates of fixed concentration of GA within a day. The inter-day precision was examined by analyzing six replicates of the same concentration of GA on different days. The results were described as % RSD and are shown in Table 2. The % RSD was within the limit of the acceptance criteria, which was not more than 2%.

Determination of LOD and LOQ

LOD and LOQ were quantitatively determined in accordance with the standard deviation method. The LOD was calculated using the formula LOD = 3.3 × SD/S, and LOD was calculated using the formula LOQ = 10 × SD / S, while SD is the standard deviation of the response and S is the slope of the calibration curves. The results were found to be 2.069 and 6.271 µg/mL, respectively Although the LOD and LOQ values in this study were higher than in the previous study [4], it was lower enough for precise and accurate quantification of GA in a gamboge sample.

Table 2. Precision results of gambogic acid.

Intra-day (n = 6)				Inter-day (n = 6)			
Conc. of GA (µg/mL)	Observed Conc. of GA (µg/mL)	Mean ± SD	% RSD	Conc. of GA (µg/mL)	Observed Conc. of GA (µg/mL)	Mean ± SD	% RSD
20	21.101	20.947 ± 0.114	0.546	20	19.907	19.527 ± 0.198	1.012
	21.025				19.568		
	21.008				19.483		
	20.889				19.398		
	20.856				19.398		
	20.805				19.407		

3.2. Determination of GA in gamboge resin

The GA content was calculated using the regression equation (y = 11.803x + 3.1421) obtained from the standard curve. The amount of GA content was 444.2 (± 0.125) mg/g or 44.42% for the determination of three different gamboge samples. The GA content in gamboge resin was reported previously about 337 mg/g [8] and 257 mg/g [4]. The data showed the content of GA was much higher in the developed HPLC method. The chromatogram of GA in the gamboge raw sample is presented in Figure 2b.

4. Conclusions

This study provided a developed and validated HPLC method for the analysis of GA in gamboge resin. The validation results exhibited that the method was simple, reliable with linearity higher than 0.9998, recovery was within 98.87–102.92%, and the RSD was not more than 2% for precision. Thus, the developed method was acceptable for the quality control of GA in gamboge resin.

Acknowledgements

The researchers are thankful to Faculty of Pharmacy (the Research and Creative Fund), Silpakorn University. The authors would like to thank Thailand Science Research and Innovation (TSRI) and Silpakorn University Research, Innovation and Creativity Administration Office for kind support and facilitation.

References

[1] Siji A and K. B. Rameshkumar 2017 *Diversity of Garcinia species in the Western Ghats: phytochemical perspective, JNTBGRI. India.* 162-169
[2] Liu Y, Chen Y, Lin L and Li H 2020 *Int. J. Nanomedicine* **15** 10385-99
[3] Zhou Z and Ma J 2019 *Exp. Ther. Med.* **18(4)** 2917-23

[4] Song JZ, Yip YK, Han QB, Qiao CF and Xu HX 2007 *J. Sep. Sci.* **30(3)** 304-9
[5] Hua X, Liang C, Dong L, Qu X and Zhao T 2015 *Biomed. Chromatogr.* **29(4)** 545-51.
[6] Qu W, Li Y, Shi D and Qu F 2015 *Se pu* **33(2)** 152-7
[7] ICH Q2(R1) 2005 Validation of analytical procedures: text and methodology, ICH Harmonised Tripartite Guideline (Chicago, USA)
[8] Khaing EM, Chantadee T, Intaraphairot T and Phaechamud T 2019 *Key. Eng. Mater.* **819** 215-20.

Physico-chemical properties of Artificial tear ducts from Fractionated Thai silk fibroin

P Limteamcharoen[1,2] and S Kanokpanont[1,2,*]

[1]Biomaterial Engineering for Medical and Health Research Unit
[2]Department of Chemical Engineering, Faculty of Engineering, Chulalongkorn University, Bangkok 10330, Thailand

*Corresponding author: sorada.k@chula.ac.th

Abstract. Artificial tear ducts (ATD) or Jones tubes, have been used in patients with tear ducts obstruction for draining tears into the nose. Commercial ATDs are generally made of glass but they have slippery surfaces causing the ducts pushed out and easily broken. Polymer-based ATDs, are more flexible and adhesive to tissues. However, the users could encounter material-related problems due to allergic reactions, low tissue adhesion and complications. We designed artificial tear ducts from Thai silk fibroin (SF) solution. The SF was fractionated using freeze-thaw cycles (-4°C and 25°C for 5 cycles) into SF-P (precipitants, 63.53±4.58% yields) and SF-S (soluble, 34.05±5.76% yields (w/v)). The concentrated solutions (20%W/W) of SF, SF-P and SF-S were dip-coated into tubes. The leaking test was performed using simulated natural tear flow rate of 0.0022 mL/min for 6 h. SF-S duct lost its shape and leaked. Absorption of the balanced salt solution (BSS) of the SF and SF-P were at 5.67±0.76% and 8.05±1.28 wt.% respectively, giving their wet inner and outer diameters at 2 and 2.5 mm. and the thickness 500 microns. Crystallinities of SF and SF-P analyzed using ATR-FTIR, were at 42.27 and 44.51% respectively. The thermal decomposition temperature of SF and SF-P analyzed using TGA, were at 277 and 280°C. Degradability in BSS containing lysozyme 1.69 U/mg to mimic tears at 37°C *in vitro* showing that both ATDs are stabilized for at least 4 weeks. BSS height obtained capillary test that compared with glass ducts, SF and SF-P at 7.0±0.0, 6.9±0.4 and 7.1±0.4 mm, respectively, while glass ducts were 34.81±0.0. The SF and SF-P have flexural stress were 55.24±7.68 MPa and 81.42±2.71 MPa, %flexural strain at max stress was 5.43±0.46% and 2.54±0.17% and flexural Modulus was 1.82±0.29 GPa and 6.13±0.51GPa respectively. The results of all experiments indicate that SF-P had the highest potential for further development into natural polymer-based ATDs.

1. Introduction

Artificial tear ducts (ATD) or Jones tubes, have been used in patients with tear ducts obstruction for draining tears into the nose. Commercial ATDs are generally made of glass but they have slippery surfaces causing the ducts to be pushed out and easily broken [1-2]. Polymer-based ATDs, such as silicone and polyethylene are more flexible and adhesive to tissues [3-4]. However, the users could encounter material-related problems due to allergic reactions, low tissue adhesion and complications. Generally, ATDs are reported to have 1.5 - 3.5 mm. inner diameter, 18-40 mm. length.

Thai silk fibroin is a natural protein derived from domesticated *Bombyx mori* that is known for its biocompatibility, slow biodegradability and good mechanical properties. Which has been reported silk fibroin scaffold for abdominal wall facial reinforcement proved to be free of complication rate and irritation in 93.5% volunteer when applied to abdomen [5], silk fibroin graft tube was implanted into the right carotid artery of a mice was reported to be stabled for 1-3 month [6]. The trilayered silk scaffolds fabricated for vascular tube had ultimate tensile strength of 12 MPa [7]. Therefore, We investigated the feasibility of using fractionated silk fibroin using freeze-thaw cycles to fabricate ATDs.

2. Experimental

2.1 Preparation of SF solution

Silk cocoons (*B. mori*, *Luengpairoj,* produced by The Queen Sirikit Department of Sericulture, Ministry of Agriculture and Cooperatives, Nakhon Ratchasima, Thailand) were degummed with 0.02 M $NaCO_3$ and dissolved in 9.3 M LiBr at 60°C for 4 h. The dissolving protein was dialyzed in DI water for 72 h and was centrifuged at 9000 rpm [8]. The finished silk solution was stored at a temperature of 4°C.

2.2 Fraction of SF Solution

The silk fibroin protein solution was diluted with water to 6%W/V (SF), then it was fractionated using freeze-thaw cycles (-4°C and 25°C for 5 cycles) into precipitants (SF-P) and soluble (SF-S).

2.3 Fabrication of artificial tear ducts (ATDs)

The protein solutions were concentrated to 20%W/W using dialysis against 20%W/V PEG (Poly ethylene glycol, MW 35,000 Da, Sigma-aldrich, USA). Then they were dip-coated to form cylindrical ducts with the inner and outer diameters of 2 and 2.5 mm., respectively, with the thickness of 0.5 mm. Beta-sheet structure of the silk protein was induced by submerging the ducts under 70% ethanol solution for 2 h then they were dried in a fume hood [8].

2.4 Characterizations

The silk fibroin solutions were characterized as follows: 1) yield (%dry weight), 2) molecular mass (SDS-PAGE), 3) viscosity (Oscillatory Rheometer (The Haak Mars Rheometer, Thermo Scientific, USA) at the shear rate of 0-100 per second, at 28°C.), 4) hydrophilicity (water contact angles of thin films that analyzed by Drop Shape Analyzer (DSA 10/MK2, Kruss, Germany)). After the protein was fabricated into ATDs. They were characterized on 1) morphology (camera and SEM (JEOL JSM-6610LV, Oxford X-Max 50, Australia) after immersion under balanced saline solution (BSS) for 24 h before freeze-drying), 2) Fourier-transform infrared spectroscopy)(FTIR, PERKIN ELMER Spectrum GX, USA) in Attenuated Total Reflectance (ATR) mode for chemical structure and percentage of crystallinity), 3) Thermal properties (thermogravimetric analysis (TGA/DSC1, Mettler Toledo, USA) at 30-800°C, heating rate of 10°C /min, N_2 gas flow rate 50 mL/min), 4) BSS absorption (weight differences after immersing under BSS for 24 h, 37°C), 5) In vitro, degradability (in BSS containing 1.69×10^5 Unit/mg lysozyme [9] to mimic natural tear at 37°C *in vitro* for 4 weeks), 6) The leaking (flow simulation of BSS through the ATDs with a flow rate of 0.0022 mL/min and a pressure of 101.52 kPa [10]), 7) Capillary action (height of in the ATDs, compared to glass ducts [11]), 8) Flexural strength (three-point bending mode, universal testing machine (Instron 5567, USA) at 1 mm/min speed, 1 kN force [12]).

All characterizations were done at 4 replications. (Except extraction yield, crystallinity index, T_D and flexural strength). Static analysis was done using Minitab program, the significant difference was determined from the 95% confidence level (P-value > 0.05).

3. Results and discussion

Silk fibroin is composed of three protein subunits: heavy-chain (350 kDa), light-chain (25kDa) and P-25 (30 kDa). SDS-PAGE of pure Thai silk fibroin (Nangnoi Srisaket) SDS-PAGE had only a single clear band at 25kDa and a smear protein at the higher 200 kDa [13]. In this research, the SF sample of Luengpairoj showed smeared bands ranging from 15 to 250 kDa (Figure 1), and the clear band at 25 kDa. The SF-P and SF-S had similar smeared band. Clearer bands at 25 kDa and lower than 20 kDa can be observed in SF-S. Properties of silk fibroin solutions were shown in Table 1. Appearance viscosities (Pa.s) of SF-P was significantly higher than those of the SF and SF-S indicated that its protein structure was larger and more extensive [14]. We can see the evidence from SDS-PAGE that the SF-S content lower molecular weight protein molecules. Water contact angles of SF films were slightly higher than those of the SF-P and SF-S. However, all of the samples had water contact angles that indicate hydrophilic materials.

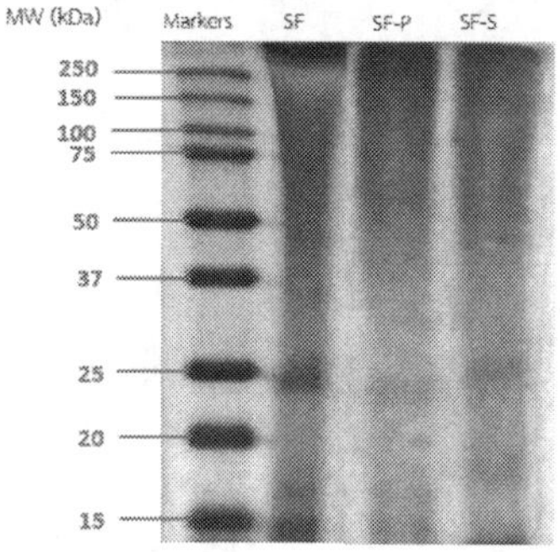

Figure 1. Molecular weight of Thai silk fibroin samples (Leung Pairoj) and their distribution compared to the standard protein makers in the SDS-PAGE analysis.

Table 1. Extracted yield of Thai silk fibroin solutions, appearance viscosity and water contact angle of solutions.

Samples	Yield of extraction (%(wt./wt.)) (n = 5)	appearance viscosity (η, Pa.s) (n = 4)	Water contact angle (°) (n=4)
SF	90.84 ± 0.87%[a]	0.47 ± 0.03[d]	67.95±3.32[g]
SF-P	63.53 ± 4.58%[b]	1.33 ± 0.42[e]	65.32±3.41[h]
SF-S	34.05 ± 5.76%[c]	0.68 ± 0.08[f]	62.39±2.17[i]

a, b, c were showed statistically significant difference at the 95% confidence level (P-value > 0.05).
d, e, f, g, h, i were showed no statistically significant difference at the 95% confidence level (P-value > 0.05).

Figure 2 showed the inner cross-section surface of the ATDs. SEM morphology revealed the SF-P had good polymer-water miscibility observed from porous structure all over the entire surface. This supported by the result from SDS-PAGE and water contact angle that the SF-P has high MW and less hydrophilicity than SF-S. SF and SF-S ATDs had leaf-like structures, showing the interaction of intermolecular polymer chains were higher than their interaction with water.

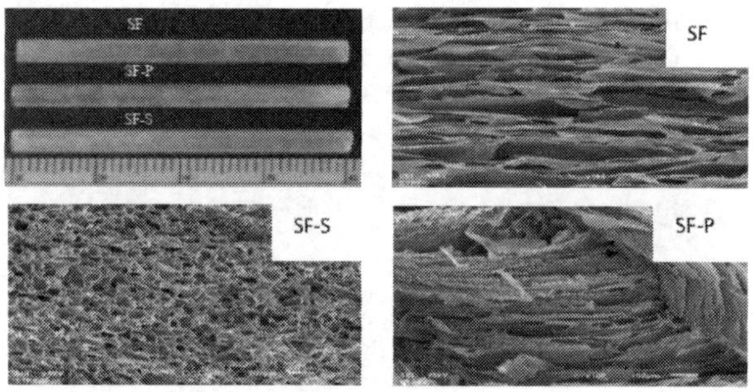

Figure 2. Morphology of ATDs, the appearance of ATDs and the cross-section surface of SF, SF-P and SF-S ATDs.

Typical peaks in ATR-FTIR spectra for silk fibroin are characteristic reported at the of the amide I (C=O stretching, 1700-1600 cm^{-1}), amide II (N–H bending, C–N stretching, 1600-1500 cm^{-1}) and amide III (N–H deformation, C–N stretching, 1350-1200 cm^{-1}) bands of proteins [15]. We found the characteristic peaks of silk fibroin at 1) 1646, 1539 and 1230 cm^{-1} (silk I, coils), 2) 1620, 1513 and 1260 cm^{-1} (silk II, β-sheet and β-turn) at the amide I, II and III, respectively (Fig.3). Crystallinity index calculated from Eq. 1 using amide I peak only due to the clear visible differences between silk I and silk II peaks [12].

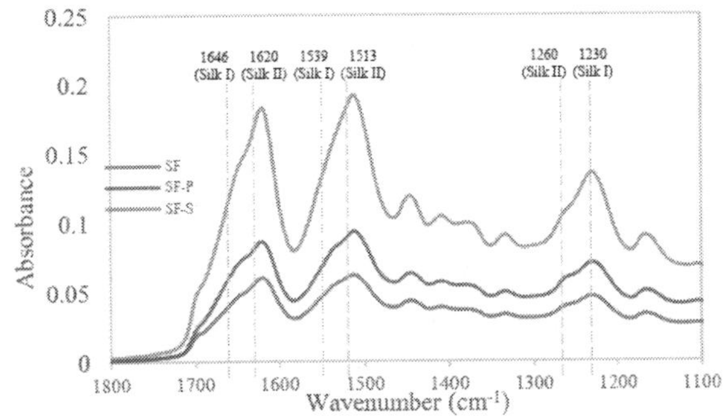

Figure 3. FTIR spectra of amide I, II and III from ATDs.

prepared from different silk fibroin samples

$$\%\text{crystallinity index} = \frac{A_{silk\,II}}{A_{silk\,II} + A_{silk\,I}} \quad (1)$$

The results from Table 2 showed that the fractionated silk fibroin had higher crystallinity indices compared to the SF indicating the higher ordered structure. These results in the higher decomposition temperature BSS absorption of SF, SF-P and SF-S ATDs were found at 5.67 ±0.76%, 8.05±1.28 and 13.18±0.33 wt.%, respectively. Their hydrophilicity was in lined with the water contact angle results. All of the ATDs loss 10 wt.% in BSS solution containing lysozyme in 4 weeks. However, SF-S had tendency to degrade faster than the other. The ATDs from SF-S failed the leaking test so those from SF and SF-P were tested further. The capillary tests of glass, SF and SF-P tube were found BSS height at 7.0±0.0, 6.9±0.4 and 7.1±0.4 mm, respectively. The results of three-point bending test showing the superior mechanical strength of the SF-P ATDs compared to those of the SF (Table 2) and the flexibility (flexural strain) of SF was higher.

Table 2. Crystallinity index (%), Thermal decomposition temperature, and flexural strength of the ATDs.

Samples	Crystallinity index (%)	T_D (onset, °C)	Ultimate flexural stress (MPa) (n=5)	Flexural strain at max stress (%) (n=5)	Flexural Modulus (GPa) (n=5)
SF	42.27	277	55.24±7.68[a]	5.43±0.46[c]	1.82±0.29[e]
SF-P	44.51	280	81.42±2.71[b]	2.54±0.17[d]	6.13±0.51[f]
SF-S	44.11	280	-	-	-

a, b, c, d, e, f were showed statistically significant difference at the 95% confidence level (P-value > 0.05).

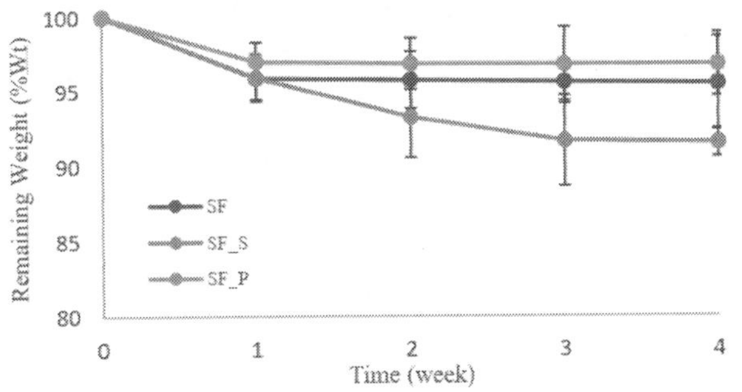

Figure 4. Percentage of Remaining weight of ATDs

The results of SF, SF-P and SF-S were showed non-statistically significant difference at the 95% confidence level (P-value > 0.05).

4. Conclusions

All physico-chemical results, especially ordered structure and mechanical properties indicated that the ATDs fabricated from Thai silk fibroin fractionated protein (SF-P) has high potential to be use as ATDs.

Further intensive studies on biocompatibility and degradation, especially the freeze-thawn induced precipitants, should investigate.

References

[1] Jones L.T. 1965 *Am J Ophthalmol* **66** 506-24.
[2] P. Komínek 2007 *Springer-Verlag Berlin Heidelberg* **1** 128-141.
[3] Eric A. Steele 2016 *Wolters Kluwer Health, Inc* **27**(5) 439–4420.
[4] Mimura M., *et al.* 2018 *American Journal of Ophthalmology Case Reports* **11** 41-44.
[5] Clemens M.W., Downey S. Agullo Frank, Lehfeldt M. R., Kind G. M., Palladino Humberto, Marshall Deirdre, Jewell M. L. Mathur A. B., Bengtson B. P. 2014 *Plastic and Reconstructive Surgery Global Open* **2**(11)
[6] Tanaka K., Fukuda D., Higashikunim Y., Hirata, Y., Komuro I., Saotome Toshiki, Yamashita Y. Asakura T., Sata M. 2020 *J. of Atherosclerosis and Thrombosis* **27**.
[7] Xili D., Tongqiang Z., Xianghui G., Changhui R., Hongyan K., Peng X., Haifeng L., Yubo F. 2016 *J. of Bioactive and Compatible Polymers* **31**(6) 613–623.
[8] Kaewprasit K., Amornrat P., Kanokpanont S., Damrongsakkul S. 2014 *App. Bio. Mat.* **102**(8)
[9] O. Yoshiki, D. Murat, T. Kazuo 2006 *Clinica Chimica Acta* **369** 17-28.
[10] S.Mishima, *et al.* 1996 *Investigative Ophthalmolog* **5**(3)
[11] Masoodi R., Ehsan L., Alireza O. 2013 *J. of Colloid and Interface Science* **389**(1) 268–272
[12] Scărlătescu D. D., Modrea A., Stanciu M. D. 2019 *Procedia Manufacturing* 32 179–186.
[13] Wadbua P., Promdonkoy B., Maensiri S., Siri S. 2010. *Int J. of Bio. Macro.* **46**(5) 493–501.
[14] Ijaz Ali & Luqman Ali Shah 2020 Polymer Bulletin 78 1275–1291
[15] Xiao H., Kaplan D., Cebe P. 2006 *Macromolecules*, **39**(18), 6161–6170.

Wood plastic composites (WPCs) from multilayer packaging wastes and rHDPE as pallets for green industry

T Boonsri[1], S Rukchonlatee[1,2], and I Jangchud[1,*]

[1] Department of Chemistry, School of Science, King Mongkut's Institute of Technology Ladkrabang (KMITL), Bangkok 10520, Thailand
[2] Polymer Synthesis and Functional Materials Research Unit, School of Science, KMITL, Bangkok 10520, Thailand

*Corresponding author: *ittipol.ja@kmitl.ac.th*

Abstract. People are becoming more aware of plastic pollution. Consumers, researchers, and manufacturers are seeking ways to reduce their contribution to the problem. Management of plastic wastes has still been severely environmental problems, especially those with multilayer or multicomponent structures due to the difficulty in recycling process. According to the concept of circular economy, this research aimed to overcome these problems by recycling the multilayer plastic film wastes after use from consumers (classified as post-consumer recycled resin, PCR) and transforming those into high-value products. In this work, Wood Plastic Composites (WPCs) made from mixtures of multilayer packaging wastes (PCRs), recycled high density polyethylene (rHDPE) and wood powders were studied in order to be used as pallets in green industry. Studied factors affecting WPC properties included PCR types (linear low density polyethylene (LLDPE), multilayered oriented polyamide (OPA)/PE and polyethylene terephthalate (PET)/PE films) and rHDPE:PCR weight ratios (40:5, 35:10, 30:15 and 25:20). WPC samples were compounded by a two-roll mill and then shaped into samples by a compression molding machine. Morphological, mechanical and thermal properties of WPCs were examined. From SEM study, it was revealed that wood powders and fibers were well distributed and randomly oriented in WPCs. Both PCRs (OPA/PE and PET/PE) were partially melted because the processing temperatures were below the crystalline melting temperatures of OPA and PET. PCR orientation in WPCs was aligned with the compression direction. Mechanical properties of WPCs from OPA/PE were similar to those of WPCs from PET/PE. WPCs from rHDPE:PCR weight ratios of 40:5 and 35:10 had good mechanical properties which passed the industrial target values. Optimum formulas were then used to form WPC samples by extrusion technique and fabricated into WPC pallets. It was found that the WPC pallets from wastes showed promising properties with high potential to replace conventional pallets in warehouses and industrial transport for green industry.

1. Introduction

Mismanagement and lost opportunities in the recycling of plastic wastes have resulted in global concerns, mostly related to widely spread plastic pollution of terrestrial and marine ecosystems [1]. The current management of plastic wastes largely relies on incineration and landfilling, which are typical technologies applied in a linear polymer economy. Recycling plastic wastes could promote economic and environmental benefits and drive the transition to a circular polymer economy [2]. Especially, polymer-based multilayer packaging materials are commonly used in order to combine the respective

IOP Conf. Series: Materials Science and Engineering 1234 (2022) 012022 doi:10.1088/1757-899X/1234/1/012022

performance of different polymers. However, because of multilayer plastic packaging wastes after use from consumers (PCRs) with their poor recyclability, most multilayers are usually incinerated or landfilled, counteracting the efforts towards a circular economy and crude oil independency [3].

Wood Plastic Composites (WPCs) have developed rapidly in recent years and are experiencing fast market growth [4]. Such products can include recycled polymers and fibers as material ingredients. WPCs also contribute to the economy by the use of recycled thermoplastics as HDPE (named as rHDPE). Using rHDPE granulated from post-consumer packaging is economically possible for WPCs. Such steps promote an increase in the value of mixed stream recycled materials and broaden the sustainability message by waste stream diversion towards a circular economy and industrial transport for green industry [5-7]. This research aimed to use mixtures of PCRs, rHDPE and wood powders for WPCs pallets. Pallets were made of wood powder:polymer with the weight ratio of 55:45. Effects of the polymer mixture of rHDPE/PCR weight ratio and PCR type were examined.

2. Experimental

2.1 Materials

Wood powders, which were screened with a 60 mesh sieve, were provided by BBJ Group Co., Ltd., Thailand. rHDPE was supplied by Iamplastic Co., Ltd., Thailand. PCRs (LLDPE, OPA/PE and PET/PE) were crushed to be about 1-5 mm length from Thai Mangkorn Plastic Industry Co., Ltd., Thailand. Calcium carbonate ($CaCO_3$: C-75(S) grade) from Chinchana Co., Ltd. Thailand was used as a chemical additive. Compatibilizers were given by Dow Chemical Company. Polyethylene wax (PE wax), used as a lubricant, was purchased from Multiple Plus Co., Ltd. Thailand. The anti-oxidants were from BASF Schweiz AG Plastic Additives 4057 Basel, Switzerland.

2.2 Compounding

Mixtures of polymers (rHDPE and PCR), wood powders and other additives (which were $CaCO_3$, PE wax, compatibilizers, anti-oxidants, proprietary information) were compounded using a two-roll mill at 190°C for 15 min with 12 rpm rotating speed. The test specimens were shaped using a compression molding machine at 190°C. The wood powder/polymer weight ratio was constantly fixed at 55:45. The WPCs formulas with various rHDPE/PCR weight ratios are presented in Table 1.

Table 1. WPCs formulas with various rHDPE/PCR weight ratios.

Sample name	**HD45**	**HD40LLDPE05**	**HD35LLDPE10**	**HD30LLDPE15**	**HD25LLDPE20**
rHDPE	45	40	35	30	25
*PCR (LLDPE)	0	5	10	15	20

*When PCR is OPA/PE or PET/PE, the sample name is HDxxPAPExx or HDxxPETPExx, respectively.

2.3 Characterization and testing

The samples were observed under scanning electron microscopy (SEM) to obtain high magnification images to show morphology of wood particles and their interaction with the thermoplastic matrix. PCRs were characterized using a thermogravimetric analyzer (TGA). Flexural properties of samples were performed according to the ASTM D790 using a universal testing machine with a test speed of 10 mm/min. Impact property of WPCs with V-notched specimens was evaluated using a Izod impact tester according to the ASTM D256.

3. Results and discussion

3.1 TGA of PCRs components

Thermogravimetric analysis of PCRs was performed under oxygen and nitrogen atmospheres. TGA results revealed that LLDPE PCR consisted of 100% PE, OPA/PE PCR was about 22% PA and 78% PE and PET/PE PCR had about 26% PET and 74% PE.

3.2 Morphology of WPCs

SEM micrographs at the 150x magnification of the cryogenically fractured surface of the WPCs from rHDPE/PCRs weight ratio of 35/10 are demonstrated in Figure 1.

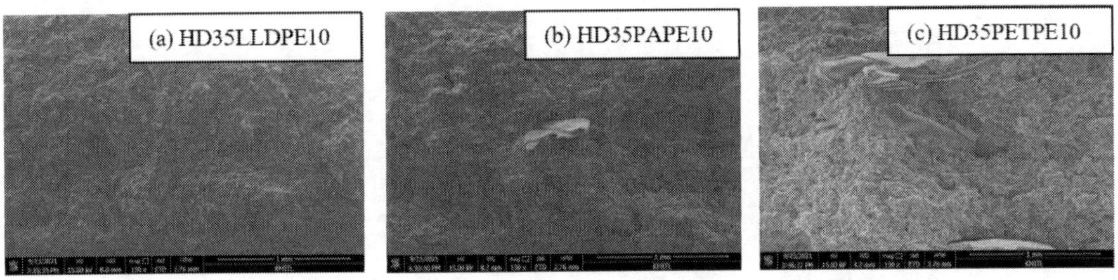

Figure 1. SEM micrographs (150x) of the fractured surface of the WPCs.

Figure 1(a) shows completely molten LLDPE PCR in the WPC matrix. However, OPA/PE and PET/PE PCRs were partially melted as seen film left in the WPC matrix in Figure 1(b-c) because the processing temperatures were below the crystalline melting temperatures of OPA and PET.

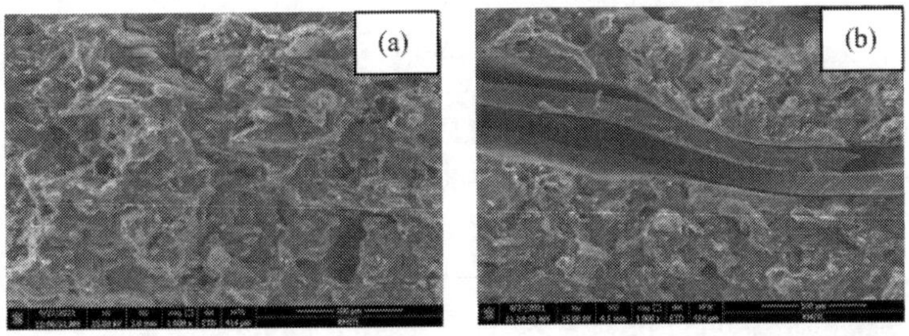

Figure 2. SEM micrographs (1000x) of the fractured surface of the HD25PAPE20 WPCs.

Figure 2 illustrates SEM micrographs at the 1000x magnification of the cryogenically fractured surface of the HD25PAPE20 WPCs. Figure 2(a) revealed that wood in forms of powders and fibers were well distributed and randomly oriented in WPCs. In Figure 2(b), it is important to consider that fault zones can influence the properties of the WPCs. These voids or fissure could be generated between the matrix and PCRs at the interface, due to weak interfacial adhesion when compared to strength of the matrix and PCRs. Therefore, voids or holes are from adhesive failure.

3.3 Mechanical properties

3.3.1 Flexural properties

Figure 3(a) shows the flexural strength (solid line) and flexural modulus (dash line) of WPCs versus PCR loading (calculated in rHDPE/PCR part). It shows that mostly the flexural strength and modulus of WPCs made from PCR loading up to 10 wt.% are similar to those of WPCs without PCR. Moreover, the WPCs incorporated with 10 wt.% PET/PE PCR has flexural modulus 23% higher than WPCs without PCR. This was caused by the unmelted PET in PCR acted as filler that could reinforce the WPCs. As the industrial target values of flexural strength and modulus of WPCs for pallets are 45 MPa and 1800 MPa, respectively, these properties of studied WPCs made from PCR loading up to 10 wt.% meet the industrial satisfaction. Nevertheless, the decreasing trend of both flexural strength and modulus of WPCs are presented when adding 15-20 wt.% PCR, regardless of PCR type. As seen in Figure 2(b), defects in the WPCs are observed due to weak interfacial adhesion between PCR (OPA/PE or PET/PE) and polymer matrix. Therefore, more defects could be taken place at higher PCR contents. In the case of LLDPE (as PCR), it was because of stronger rHDPE as compared with LLDPE.

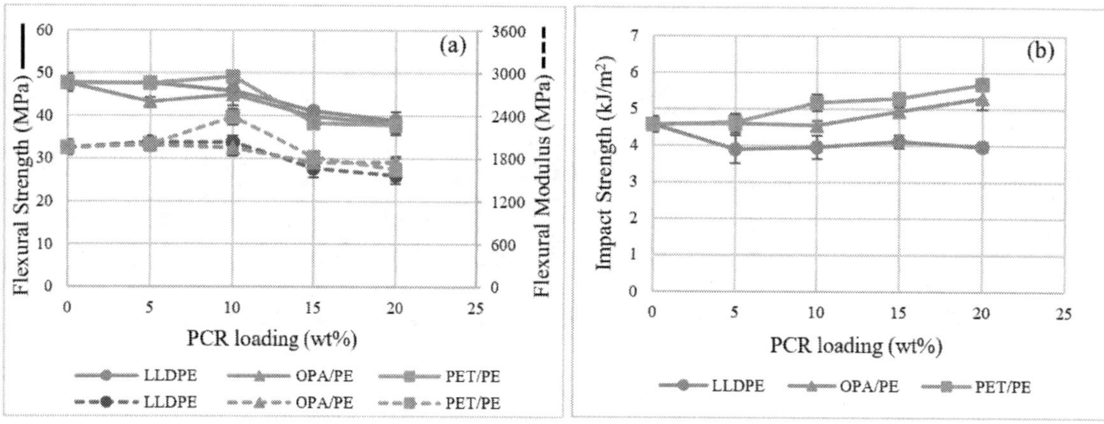

Figure 3. Flexural properties and impact strength of WPCs.

3.3.2 Impact strength

Figure 3(b) demonstrates the impact strength of WPCs versus PCR loading (calculated in rHDPE/PCR part). An increase in PCR loading (OPA/PE and PET/PE) increases impact strength of WPCs. This could be contributed to the specimen preparation technique (compression process), resulting in the alignment of unmelted OPA and PET (as PCRs) with the compression direction. Therefore, as the amount of PCRs increased, higher impact force transfer and greater absorbed energy were occurred.

4. Conclusion

With the WPCs having the wood powder:polymer weight ratio of 55:45, the use of polymer having rHDPE:PCR weight ratios 40:5 and 35:10 gave good mechanical properties, passing the industrial target values of WPC pallets with regardless of the PCR type studied. This research revealed that PCR, rHDPE and wood wastes could be used as alternative raw materials in the manufacture of WPCs. The designed formulas resulted in sustainable production of WPCs pallets at a lower cost with high potential to replace conventional pallets in warehouses and industrial transport for green industry.

Acknowledgement

Authors would like to thanks MBJ enterprise Co., Ltd. for support materials and laboratory. Many thanks to Dow Chemical Company for chemicals and other supports. The financial support from Department of Chemistry, School of Science, King Mongkut's Institute of Technology Ladkrabang (KMITL) has also been greatly appreciated.

References

[1] Victor S, Paulo F, Keith L and Grey W 2021 *Polym. Degrad. Stab.* **190** 109627.
[2] Jijiang H, Andrei V, Wei Ping C, Apostolos G and Grzegorz L 2022 *Renew. Sust. Energ. Rev.* **125** 111866.
[3] Kaiser K, Schmid M and Schlummer M 2017 *Recycl.* **1** 3010001.
[4] Zhou Y, Stanchev P, Katsou E, Awad S and Fan M 2019 *J. Waste Manag.* **19** 42-48.
[5] Philipp F, Johannes W and Andreas K 2015 *J. Waste Manag.* **46** 76-85.

[6] Yonny L, Juarez P, Donatian G and Fabricio G 2020 *J. Clean. Prod.* **244** 118723.
[7] Leu S, Yang T and Lo S 2012 *Constr. Build. Mater.* **29** 120-127.

IOP Conf. Series: Materials Science and Engineering 1234 (2022) 012023 doi:10.1088/1757-899X/1234/1/012023

Characterization of injection molded polylactide stereocomplexes blended with thermoplastic starch and chain extender

Y Srithep[1,*], D Pholharn[2], O Veang-in[1] and J Morris[3]

[1]Manufactuing and Materials Research Unit, Faculty of Engineering Mahasarakham University, Mahasarakham 44150, Thailand
[2]Department of Chemistry, Faculty of Science and Technology, Rajabhat Mahasarakham University, Mahasarakham 44000, Thailand
[3]School of Industrial Education and Technology, King Mongkut's Institute of Technology Ladkrabang, Bangkok, 10520, Thailand

Corresponding author: yottha.s@msu.ac.th

Abstract. Polylactide (PLA) is an interesting biodegradable polymer but has limited application because of its low thermal stability and brittleness. Both weaknesses of PLA were solved by stereocomplexes (ST) blended with thermoplastic starch (TPS). In this work, 15% and 30% TPS and 2% chain extender were blended with poly(L-lactide) (PLLA)/poly(D-lactide) (PDLA) (50/50) in an internal mixer. The blended materials were then injection molded into tensile specimens. Differential scanning calorimetry demonstrated that despite the added starch, the stereocomplex structures were still formed and stereocomplex crystallinity decreased with increasing starch content. Compared to neat PLLA, the melting temperature of the stereocomplex and its starch blends was about 55°C higher. The starch enhanced the elongation at break of the stereocomplex samples. Furthermore, stereocomplex and starch compatibility was improved by extending the chains, which in turn improved the stereocomplex and starch blend mechanical properties.

1. Introduction

Derived from natural sources, polylactide (PLA) also degrades naturally and has good optical, mechanical and barrier properties [1] and two enantiomers, poly(L-lactide) (PLLA) and poly(D-lactide) (PDLA). To enhance mechanical and thermal properties and resistance to hydrolysis PLA-based materials, PLLA and PDLA can be blended leading to a stereocomplex, with a 220°C melting temperature or ~50°C greater than either homocrystallized enantiomer [2,3]. The molecular weight, optical purity and chain structure (linear or branched) affect the degree of the stereocomplex formation [4].

This study aimed to improve thermal stability and toughness of PLA, hypothesizing that the stereocomplex would enhance thermal stability and thermoplastic starch (TPS) would improve the toughness. Further, it confirmed that adding chain extenders (CE) is an effective way to improve the compatibility of the blended materials. Equal quantities of PLLA and PDLA, with varying amounts of added starch, were melt blended and injection moulded. In the blend, stereocomplexes were formed, which significantly altered the PLLA properties, already studied with added starch [5]. Differential scanning calorimetry was used to evaluate and measure stereocomplex formation in the blends. Moreover, mechanical properties and morphologies were studied.

2. Experiments

2.1 Materials
PLLA (trade name L175) and PDLA (trade name D120) was obtained from Total Corbion PLA (Thailand) Ltd. Native rice starch was purchased from Thai Flour Industry, Thailand. Glycerin (Green Global Chemical, Thailand, purity 99.5%) was used as a plasticizer. To join the chains, we used a styrene acrylic copolymer (oligomeric coupling agent), equipped with epoxy functional groups (BASF Corporation Joncryl ADR4370).

2.2 Processing
For TPS preparation, native rice starch and glycerin (content 25wt%) were manually mixed and stored at 25 ± 2°C for 24 h. The resulting blends were mixed in rheometer or mixer (HAAKE Polylab OS system) at 60 rpm and 200°C for 4 min.

PLLA or ST was then mixed with 15 or 30wt% TPS with the 0 and 2wt% chain extender using the same mixer at 60 rpm and 200°C for 4 min. Tensile bars were formed in an injection molding machine (ING-58T, Chareon Tut Co., Ltd) following ASTM D638 Type I. The neat stereocomplex (PLLA: PDLA 50:50), neat PLLA and TPS were injection molded and used as reference materials.

2.3 Material Characterization

2.3.1 Differential Scanning Calorimetry (DSC). Melting temperatures of the homo, T_{m1}, and stereo-complexes, T_{m2}, and enthalpies of both structures were determined with a differential scanning calorimeter (Perkin–Elmer DSC4000, Massachusetts, USA). Samples were heated at 10°C/min from 0 to 250°C in an N_2 gas flow.

2.3.2 Scanning Electron Microscopy (SEM). The samples were frozen in liquid N_2 and then quickly impact fractured, after which an (≈20 nm) Au layer was sputter-coated on them. SEM images of the fracture surfaces were acquired with a Hitachi SU-1500 electron microscope, operated at 3 kV.

2.3.3 Mechanical Testing. Tensile properties of the injection molded specimens were measured with an NRI-TS501 universal testing instrument at a constant 10 mm/min crosshead speed.

3. Results and Discussions

3.1 Thermal Properties
Figure 1. shows DSC thermograms of PLLA blended with TPS. PLA had three thermal steps: (1) a glass transition temperature (T_g) ~60°C, (2) a cold crystallization step in the range 95-120°C, and (3) an endothermic fusion step (melting peak, T_m) with a maximum at 155-175°C. On the other hand, TPS is an amorphous polymer. When PLLA and TPS were mixed, the melting enthalpy decreased with increasing TPS content (see Figure 1). Thus, dispersing TPS into the blends reduced the crystallization. When the chain extender was added to the blends, the resulting longer chains decreased PLLA chain mobility and melting enthalpy of the blended materials.

Figure 2. shows the DSC thermogram of the stereocomplex blends (PLLA: PDLA 1:1) with varying amounts of TPS. The blended materials had a similar glass transition temperature (T_g) ~60°C but the endothermic peaks of all blends, observed from 208 to 230°C, were assigned to the melting of

IOP Conf. Series: Materials Science and Engineering 1234 (2022) 012023 doi:10.1088/1757-899X/1234/1/012023

stereocomplex crystallites, which were ≈50°C higher than that of PLLA blends, confirming almost complete formation of stereocomplex crystallites.

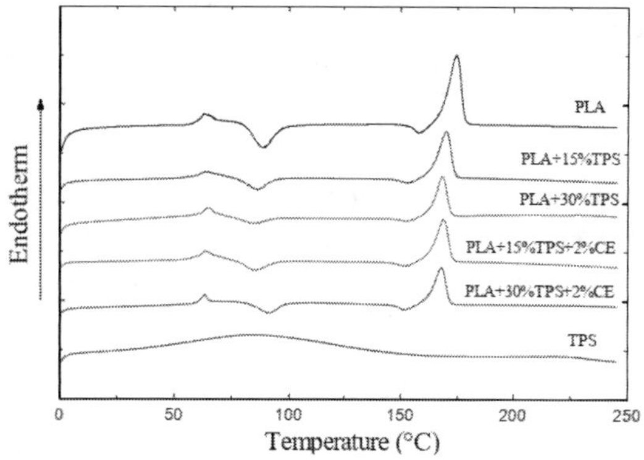

Figure 1. Thermograms of PLLA and TPS blends with CE contents, 0 and 2wt.%.

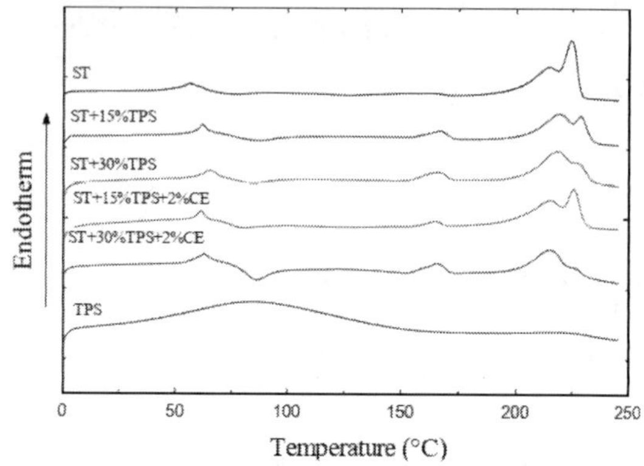

Figure 2. Thermograms of ST and TPS blends with CE content 0 and 2wt.%.

3.2 Morphology

Since the morphology of the PLLA/TPS/CE blends and the ST/TPS/CE blends was similar, only those of ST/TPS/CE blends are shown. SEM images of the stereocomplex blends are in Figure 3. They reveal the microstructure and the fractures of the specimens. The fracture surfaces of the stereocomplex and starch were relatively smooth and consistent with brittle fracture following freezing (Figure 3a and 3b), whereas the ST+15%TPS (Figure 3(c)) and ST+30%TPS (Figure 3(d)) clearly showed distinct phases. With chain extended added, compatibility between the stereocomplex and the starch TPS improved and the dispersed phase size reduced (Figure 3(e) ST+15%TPS+2%CE and (f) ST+15%TPS+2%CE). Similar behaviour was reported in blending PLLA/TPS and chain extender was used as compatibilizer [6].

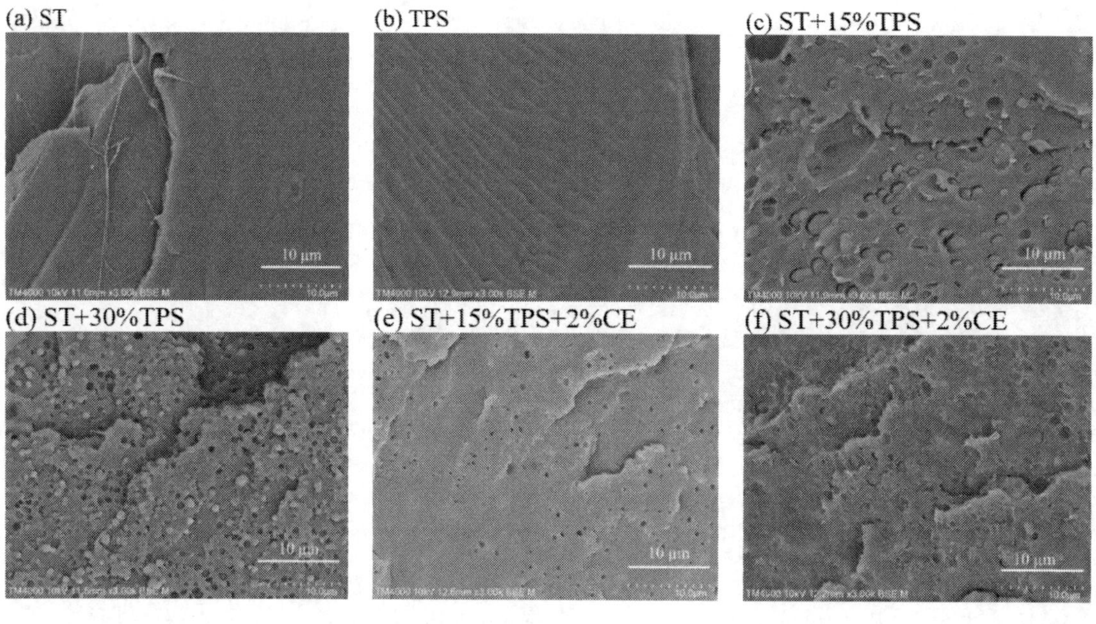

Figure 3. Scanning electron micrographs: raw stereocomplex (a) ST and starch (b) TPS; sterecomplex with added starch (c) ST+15%TPS (d) ST+30%TPS; and with added chain extender (e) ST+15%TPS+2%CE and (f) ST+30%TPS+2%CE.

3.3 Tensile Properties

Stress-strain curves were acquired for PLLA (Figure 4.) and stereocomplex (Figure 5.) specimens; blended with starch and chain extender. Figure 4 shows that the PLLA tensile stress decreased but strain-at-break increased with starch content, indicating the toughening effect of starch in the PLLA. Przybytek *et al.* (2018) incorporated thermoplastic starch (TPS) in the PLLA matrix and found that it increased the flexibility and reduced the tensile strength [5]. In our study, the chain extender enhanced this effect: elongation at break steadily increased with added chain extender, reaching 46%.

The influence of the starch on mechanical properties of stereocomplex specimens is shown in Figure 5. Adding both starch and chain extended led to tougher specimens: the strain for the stereocomplex with 30 wt.% starch increased from 1% to 10%. However, the tensile strength and elongation-at-break of stereocomplex specimens was slightly lower than those of PLLA samples. We believe that a higher density of links between chains for the stereocomplex samples led in shrinkage and warp and caused lower mechanical properties.

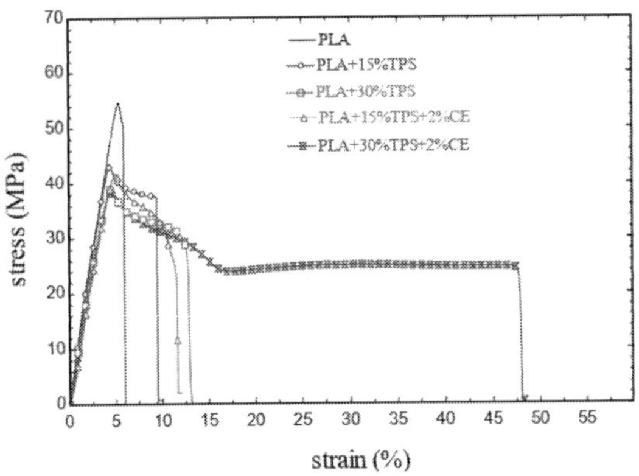

Figure 4. Stress-strain curves of PLLA and starch blends with CE content of 0 and 2wt.%.

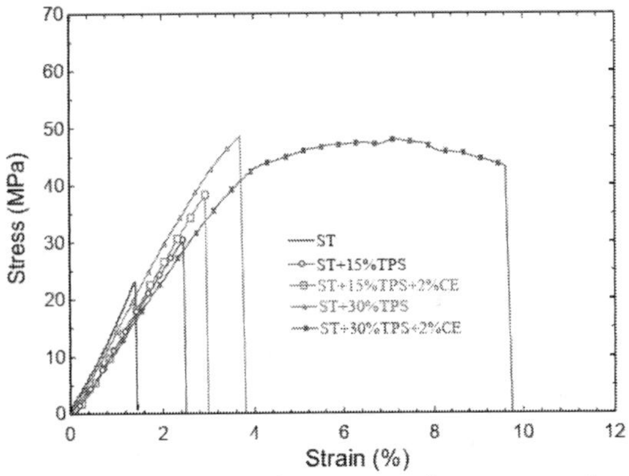

Figure 5. Stress-strain curves of stereocomplex and starch blends with CE content of 0 and 2wt.%.

4. Conclusions

In this article, the PLLA and PDLA with various amounts of thermoplastic starch were injection molded. The PLLA/PLDA blends almost completely formed stereocomplex crystallites - verified by DSC. We showed that adding the starch to the steroeomplex PLA improved toughness and expanded potential applications of stereocomplex PLA. The chain extender improved compatibility between stereocomplex and the starch and further improved the mechanical properties of the blended materials.

Acknowledgement

This research was financially supported by Faculty of Engineering, Mahasarakham University (Grant year 2022)

5. References

[1] S. Saeidlou, M. A. Huneault, H. Li, P. Sammut, and C. B. Park, Polymer, 53(25): (2012).
[2] Y. Srithep, D. Pholharn, L.-S. Turng, and O. Veang-in, Polym. Degrad. Stabil., 120: (2015).

IOP Conf. Series: Materials Science and Engineering 1234 (2022) 012023 doi:10.1088/1757-899X/1234/1/012023

[3] Y. Liu, J. Sun, X. Bian, L. Feng, S. Xiang, B. Sun, Z. Chen, G. Li, and X. Chen, Polym. Degrad. Stab., 98(4): (2013).
[4] S. Saeidlou, M. A. Huneault, H. Li, and C. B. Park, Polymer, 54(21): (2013).
[5] A. Przybytek, M. Sienkiewicz, J. Kucińska-Lipka, and H. Janik, Ind. Crops. Prod., 122: (2018).
[6] H. Li and M. A. Huneault, J. Appl. Polym. Sci., 122(1): (2011).

IOP Conf. Series: Materials Science and Engineering 1234 (2022) 012024 doi:10.1088/1757-899X/1234/1/012024

Study on rapid measurement technic of dry rubber content in latex cup lump by electrical properties

P Chaisrichonlathan[1], and C Chavapradit[1]

[1]Agricultural Engineering Research Institute, Department of Agriculture
Bangkok 10900, Thailand 10900

*Corresponding author: jasmine.1100@gmail.com

Abstract. Dry rubber content meter is substantial tool for Impartial trade on rubber cup lump. Department of Agriculture researchers proposed rapid measurement technic of dry rubber content in latex cup lumps by their electrical properties. Two probe types needle and roller were designed, in the first phase, constructed, installed and tested. Needle type probe of 25 mm. in length and 5 mm. apart was selected in measuring electrical properties of cup lump samples as its penetration depth is half of the thickness. Two cylindrical rollers probe of 220 mm. in length and 101.6 mm. (4 inches) in diameter powered by 2 HP gear motor were designed, constructed, installed and tested. Rubber cup lump samples were each statistical divided into 9 regularly measuring parts in needle type probe. Rubber cup lump samples in rollers type probe were typically enforced to pass through 1-inch gap between two rollers. Irregular size sample of more than 76.2 mm. (3 inches) thickness of rubber cup were sliced before measuring into 2 equal thickness pieces. Electrical properties of samples were acquired by standard electrical measuring instruments and percentage of dry rubber content were determined by standard oven method for all sample segments. As the results of this study, rubber content percentage of cup lump was correlated with electrical capacitance than resistance. Electrical capacity value of the latex cup lump from needle-type and two roller-type probes were inversely correlated with the percentage of dry rubber content within the range of 0 – 75 nF and 0 – 380 nF and coefficient of determinations were 0.90 and 0.97, respectively. These initiate features will be utilized in designing of percentage of dry rubber content prototype meter by electrical property values in the future. Percentage of dry rubber content from DRC method of lump from fresh latex and conventional cup lump by oven method was significantly correlate. Electrical properties of latex cup lump produced in the laboratory and farmers are also significantly consistent.

1. Introduction

Latex cup lumps are products from the upstream rubber industry. All latex cup crumbs are processed into crepe rubber by partially dehydrating. Crepe rubber will be further processed into 2 types such as compound and block rubbers. Sixty-billion-baht value or 82.8 percent was export market share of Thailand's block rubber industry [1].

Dry Rubber Content (DRC) of latex cup lump is necessary consistent to control rubber quality in each industry. But currently, trading of latex cup lump relies mainly on subjective domain. Visual assessment was used for considering the latex cup lump characteristics. There are no scientifically clear gauge tools. The latex cup lump value was deducted according to the humidity depend on considering of assessor, which bias in the quality evaluation was not able to avoid [2].

Rubber farmers attained lower selling price of latex cup lumps due to unjust subjective domain trading. Urgent request ensued on measure tool for percentage of dry rubber content in the latex cup lumps that provides accurate measurement results, fast and easy to use. Department of Agricultural proposed rapid measuring dry rubber content in latex cup lumps technique by electrical properties.

2. Experimental

2.1 Experimentation on Probe types

Two types of probes, needle and two rollers were designed and built. The needle probe was 25 mm. in length of 25 mm. and 5 mm. in distance between two needles that is approximately half the thickness of the latex cup lump. Two roller probes, each roller was 220 mm. in length and 101.6 mm. (4 inches) in diameter powered by 2 HP gear motor which determined from properties of pressure on the latex cup lumps. Latex cup lumps produced by farmers were used to be samples. In needle probe experiment, sample of rubber lump was divided into 9 parts of measuring (Figure 1). In two roller probe experiment, distance between the surfaces of two rollers was set at 25.4 mm. (1 inch) for normal thickness cup lumps (Figure 2). Sample that was thicker than 76.2 mm. (3 inches) would be cut into 2 by thickness. Measure electrical properties was determined by DT-9205A. standard electrical meter. The percentage of dry rubber contents were analyzed by drying of small pieces method at the middle position of all segments.

Figure 1. Investigation of electrical properties using needle-type probe.

Figure 2. Investigation of electrical properties using two roller-type probe.

2.2 Comparative experiment for determination of dry rubber content by DRC measurement in fresh latex method and drying of small pieces method

Fresh latexes were collected by the formulation of one latex cup lump per rubber plant was collected in plastic bottle. Dry rubber content percentage were analyzed by DRC method in the fresh latexes (Figure 3) and latex cup lumps were produced in laboratory (Figure 4). Latex cup lumps sample were cut into small pieces to find the dry rubber content percentage by drying of small pieces method. The percentage of dry rubber content was compared by DRC measurement in fresh latex method and drying of small pieces method.

Figure 3. Drying the rubber samples in DRC measurement in fresh latex method.

Figure 4. latex cup lump samples at completely settle state.

3. Results and discussion

3.1 Probe types Experimentation results

Rubber content percentage of cup lump was correlated with electrical capacitance than resistance. Electrical capacity value of the latex cup lump from needle-type and two roller-type probes were inversely correlated with the percentage of dry rubber content within the range of 0 – 75 nF and 0 – 380 nF and coefficient of determinations were 0.90 and 0.97, respectively. Detection limit of needle-type and two roller-type probes for the percentage of dry rubber content (%DRC) measurement were 55 – 84 and 54 – 76, respectively. These initiate features will be utilized in designing of percentage of dry rubber content prototype meter by electrical property values in the future.

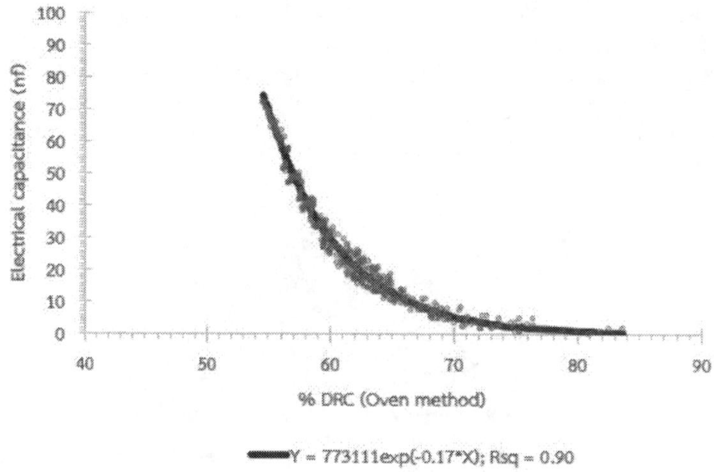

Figure 5. Relationship between electrical capacitance by measuring needle-type probe and percentage of dry rubber content.

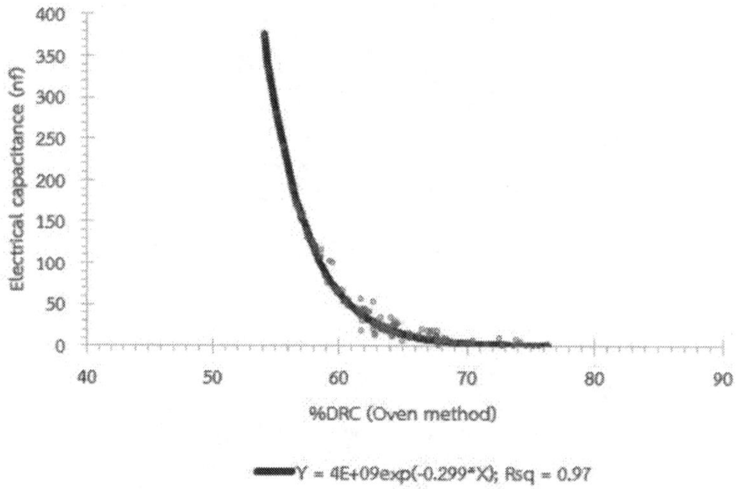

Figure 6. Relationship between electrical capacitance by measuring two roller-type probe and percentage of dry rubber content.

3.2 Comparative experiment result for determination of dry rubber content by DRC measurement in fresh latex method and drying method.

Result showed that the analytical methods for the percentage of dry rubber content by DRC measurement in the fresh latex method was high linearly correlated with drying of small rubber piece method (Figure 7).

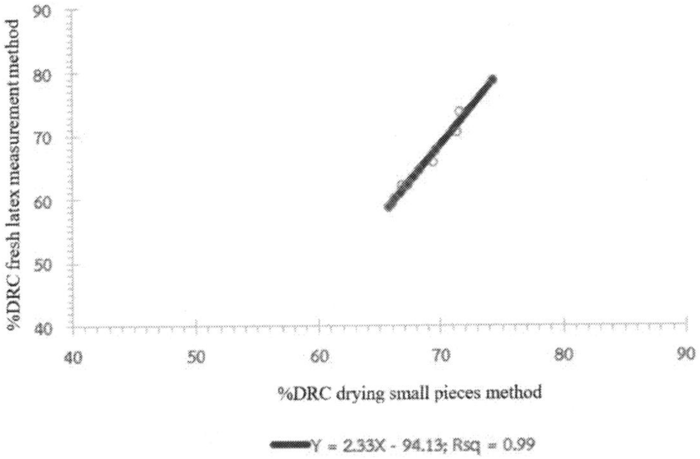

Figure 7. Relationship between dry rubber content percentage of latex cup lump between fresh latex measurement method and drying small pieces method.

4. Conclusions

Electrical capacitance properties will be utilized in designing of percentage of dry rubber content prototype meter by electrical property values in the future. Electrical capacity value of latex cup lump when using a needle probe and two roller probe were inversely correlated with the percentage of dry rubber content. For convenience of artificialing prototype meter experiment, drying of small pieces method can be used instead of the percentage of dry rubber content by DRC measurement in fresh latex method.

Acknowledgement

We would like to thank the Office of the National Research Council of Thailand (NRCT) and the Thailand Research Fund (TRF) for supporting research grants funding.

Reference

[1] Sowcharoensuk C 2021 Business/Industry Outlook 2021-2023: Processed Rubber Industry. https://www.krungsri.com/th/research/industry/industry-outlook/Agriculture/Rubber/IO/io-rubber-21. October 2021

[2] Ananta M, Sirifonnukul S and Sikarinmas S 2010 Trading pattern of rubber cup lumps in the Northeast. *Full year Research report of the 2010*, 574 – **596.** Rubber Research Institute. Bangkok

IOP Conf. Series: Materials Science and Engineering 1234 (2022) 012025 doi:10.1088/1757-899X/1234/1/012025

Properties of hydroxyapatite based geopolymer synthesized from bituminous fly ash

S Yoolamnan[1*], S Asavapisit[1*], R Piyapanuwat[2]

[1] Environment Technology Program, School of Energy, Environment and materials, King Mongkut's University of Technology Thonburi, Bangkok, Thailand.
[2] Innovative Environmental Management and Smart Construction Material Laboratory, King Mongkut's University of Technology Thonburi (Ratchaburi Learning Park), Rang Bua, Chom Bueng, Ratchaburi, Thailand.

*Corresponding author: sirirat.95@mail.kmutt.ac.th, suwimol.asa@kmutt.ac.th

Abstract This research studied the engineering properties of hydroxyapatite based on fly ash-geopolymer as a biomaterial. The ratio of Na_2O/SiO_2 and K_2O/SiO_2 at 0.15 0.20 0.25 0.30 by mole, 2.5 5.0 7.5, 10 wt.% of calcium phosphate and heat treatment at 500°C, 600°C and 700°C for 2 hours were used to studied changed of compressive strength and crystalline phase in samples. The results observed highest 28-days compressive strength of 21.99 and 19.14 MPa for 0.20 by mole of Na_2O/SiO_2 and 0.15 by mole of K_2O/SiO_2, respectively. When calcium phosphate (CP) was added in fly ash-geopolymer, the optimum ratio of CP was found in 2.5 wt.% that have unconfined compressive strength (UCS) of 19.03 and 13.60 MPa for NaOH and KOH activator. After heat treatment of geopolymer at 500°C, the maximum compressive strength at 9.09 MPa and 22.76 MPa for each activator was obtained. Finally, the XRD resulted show hydroxyapatite, mullite quartz and sodium aluminum silicate peaks in both fly ash-geopolymer with NaOH and KOH and 2.5% CP.

1. Introduction
The development of biomaterial as bone implant must consider bioactivity and biocompatibility with bone and blood in the human body. Biomaterials which were used in human such as bio-ceramic, TiO_2, composite materials. The most popular bio-ceramic material being used is hydroxyapatite (HA). It is belonging to CP compound that is quite similar to bone. Human bone consists of 60% inorganic component (or HA), 30% organic component and 10% water relative to the weight of bone [1]. HA can be synthesized using various methods, including dry, wet, thermal, and combination. Several factors affecting the synthesis of HA including pH, temperature, and the molar ratio of chemical. Generally, HA has a Ca/P ratio around 1.667 [2]. The result of each method produced a different size, morphologies, and crystalline phases of the products. Even if there are various methods to synthesise HA, the materials needed for the synthesis were too complicated [3].

Geopolymer is an aluminosilicate compound that was formed by the reactions between silica (SiO_2) and alumina (Al_2O_3) in the presence of alkaline activators such as sodium hydroxide (NaOH) and potassium hydroxide (KOH) [4]. Geopolymers have high strength similar to concrete but less CO_2 emission in production process. Mackenzie *et al.* [5] and Herve *et al.* [6] synthesized geopolymer and use as supporting material for the formation of HA. The geopolymer cement with HA has the compressive strength greater than 40 MPa that can be used as a bone implant. Therefore, the aim of this research was to synthesize HA-based geopolymer from bituminous fly ash The performance of HA-based geopolymer was assessed through the development of UCS and microstructures.

2. Experimental

2.1. Materials and Chemicals

The fly ash (FA) was used for the synthesis of geopolymer. FA from BLPC power company in Rayong province, Thailand has chemical compositions as determined by XRF is shown in Table 1. SiO_2, Al_2O_3, CaO and Fe_2O_3 were the major elemental oxide contents of FA and were present at 53.6, 25.4, 8.25 and 5.89%, respectively. XRD result in Figure 1. shows the appearance of the crystalline phases of quartz and mullite with the presence of amorphous phase between 18 and 28 degrees of 2 theta. Commercial grade chemicals in this research including NaOH from Formosoda-p, Taiwan, KOH from UNID, Korea and CP from FDSCL, Mumbai having purity of 99, 95 and 99%, respectively were used.

Table 1. Chemical composition of FA (wt%)

Chemical composition	SiO_2	Al_2O_3	CaO	Fe_2O_3	TiO_2	SiO_2/Al_2O_3
% by weight of FA	53.6	25.4	8.25	5.89	1.63	2.1

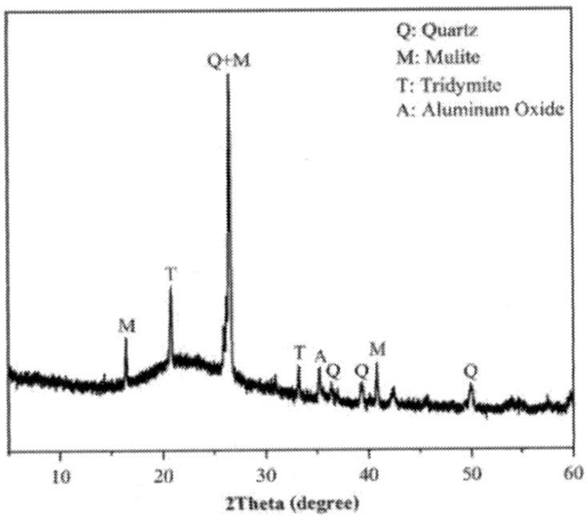

Figure 1. XRD diffractogram of FA

2.2. Preparation and testing

2.2.1. Preparation HA-geopolymer paste

The geopolymer was synthesized from FA and alkaline activators (NaOH or KOH) with a ratio of Na_2O/SiO_2 and K_2O/SiO_2 of 0.15, 0.2, 0.25, and 0.3 by mole. All materials were mixed with a mechanical mixer, following ASTM D1633-17. The mixtures were placed in cylindrical polyvinylchloride (PVC) molds with a 35 mm diameter and a 70 mm height. All samples were demolded after 24 h curing, stored at room temperature and kept for 7, 14, and 28 days prior the strength measurement. The proportion having maximum UCS was selected for the addition of 2.5, 5, 7.5, and 10 wt.% of CP and following heat treatment at 500, 600, and 700°C for 2 h

2.2.2. Method for testing

The mechanical property of all samples was measured by compaction following ASTM D1633-07 [7]. Each mix proportion was prepared for 6 replications and to be used for UCS determination. Strength results were presented with 95% confident level. In addition, the selected proportions of the cracked samples from UCS testing were kept for microstructural study by XRD technique.

3. Results and discussion

3.1 Effect of the alkaline activators on compressive strength of FA-based geopolymer.

The UCS of FA-based geopolymer as a function of increasing alkaline concentration and curing time are shown in Figure 2 (a, b). UCS increased with increasing curing time at all concentrations of both alkali activators. The highest UCS at the age of 28 days was obtained from samples with 0.2 Na_2O/SiO_2 and 0.15 K_2O/SiO_2 having strength of 21.99 and 19.14 MPa, respectively. The highly alkaline of both activators (NaOH or KOH) dissolved silica (SiO_2) and alumina (Al_2O_3) from the surface of FA particles. The soluble silica and alumina species then react with Na^+ or K^+ cations from alkali activators and transform into a tetrahedral framework of sodium or potassium aluminosilicate [8,9]. This resulting in the development of strength of the FA-based geopolymer.

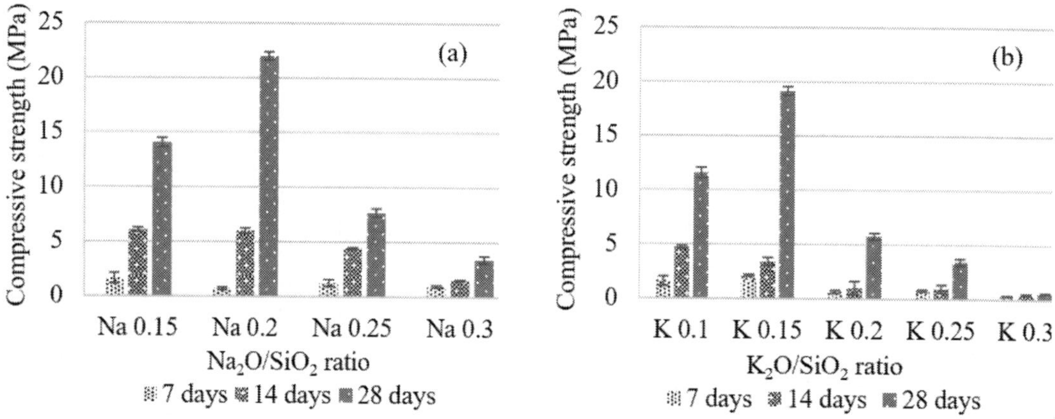

Figure 2. UCS of FA-based geopolymer with activators of (a) NaOH and (b) KOH

At lower alkali concentration than the optimum value, lower amounts of silica and alumina were dissolved from FA. Therefore, lower geopolymer products are formed and leading to lower strength On the other hand, at higher alkali concentration of both alkali activators than the optimum value, lower strength of geopolymers were gained (Figure 2 and 3). This phenomenon was caused by the highly alkali environment of high concentration activators accelerates the dissolution of silica and alumina from FA. Another explanation could be caused by a higher viscosity and temperature of the mixtures with high concentration of alkali activators leading to a rapid precipitation of geopolymer on the surface of FA particles. This could obstruct the geopolymerization reaction and thus lower strength was observed [8]. From this experiment, the optimum concentrations of NaOH and KOH at 0.2 Na_2O/SiO_2 and 0.15 K_2O/SiO_2 were selected for further study.

3.2 Effect of Calcium phosphate on compressive strength of HA-based geopolymer

A 28-days strength of FA-based geopolymer activated with NaOH varied from 21.99 MPa with no addition of CP to 19.03, 17.92 and 17.78 MPa with the addition of 2.5, 5.0 and 7.5 wt.% of CP, respectively (Figure 3 (a)). Similar variations of strength were obtained for FA-based geopolymer activated with KOH but lower strength was gained (19.14 MPa with no addition of CP to 13.60, 14.01 and 14.00 MPa with the addition of 2.5, 5.0 and 7.5 wt.% of CP, respectively) as is shown in Figure 3(b). Addition of CP causes an imbalance of ionic species in the mixtures due to Ca^{2+} and PO_4^{3-} which interfering the normal geopolymerization reactions resulting in less formation of geopolymer products thus leading to lower strength of FA-based geopolymer [6]. From the experimental results, the concentration of 2.5 wt.% CP was selected for FA-based geopolymer activated with both NaOH and KOH to study the heat treatment on strength of HA-based geopolymer.

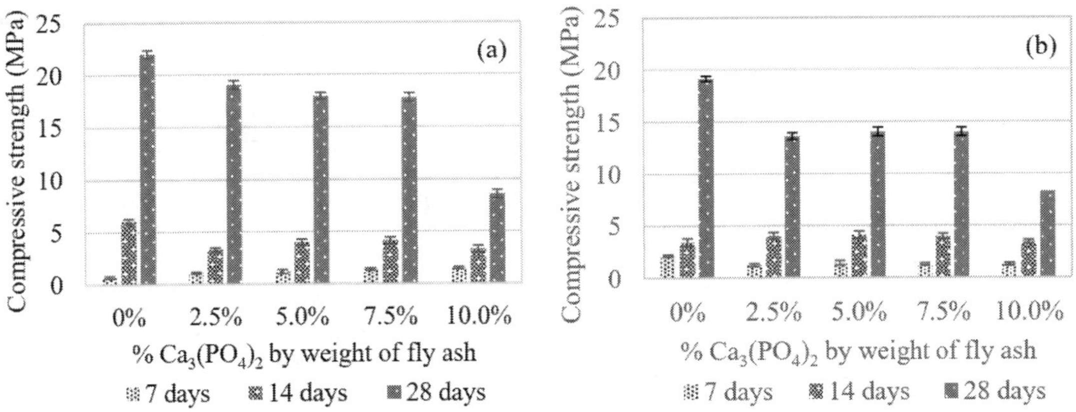

Figure 3. UCS of FA-based geopolymer activated with activators of (a) NaOH or (b) KOH and (c) NaOH or (d) KOH at various amounts of $Ca_3(PO_4)_2$

3.3 Effect of temperatures on compressive strength of HA-based geopolymer
Figure 4 shows a strength of HA-based geopolymer with both activations and 2.5 wt.% of $Ca_3(PO_4)_2$ addition after treatment at 500 600 and 700°C gain strength of 9.09, 7.37 and 5.74 MPa for samples with NaOH and 22.76, 15.82 and 20.18 MPa for samples with KOH, respectively. The highest strength was obtained from samples treated at 500°C. A continue reduction of strength for samples activated with KOH was observed after treatment at the temperature at 600°C and 700°C. This could be caused by the presence of an increase micro-cracking with increasing treatment temperature [5, 10, 11].

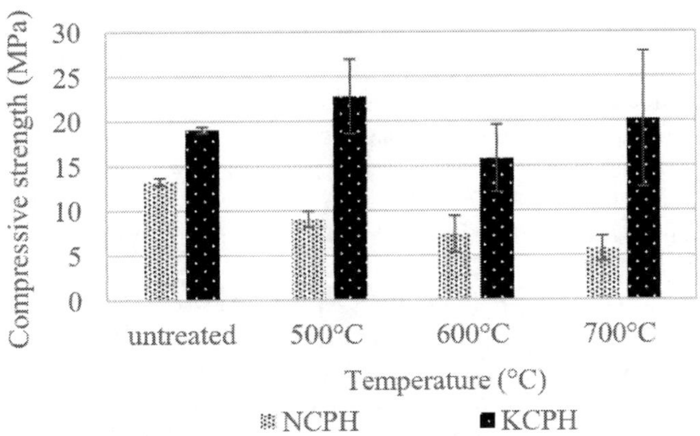

Figure 4. UCS of HA-based geopolymer after treatment at 500, 600 and 700°C

3.4 XRD results
Figures 5. a-d show XRD patterns of HA-based geopolymer activated with 0.2 of Na_2O/SiO_2 or 0.15 of K_2O/SiO_2 in the presence of 2.5% of calcium phosphate. XRD diffractograms reveals the presence of the crystalline phases including HA, mullite and quartz in all samples except for the samples activated with NaOH also found sodium aluminum silicate and KOH found potassium aluminum silicate. This is in agreement with the previous research [5,6].

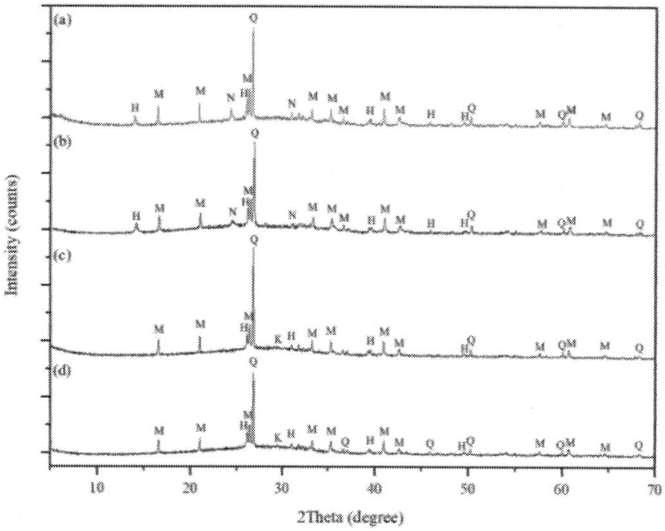

Figure 5. XRD diffractograms of (a) HA-based geopolymer with NaOH, (b) HA-based geopolymer with NaOH and treated at 500°C, (c) HA-based geopolymer with KOH and (d) HA-based geopolymer with KOH treated at 500°C

M: mullite,
Q: quartz,
H: hydroxyapatite,
N: sodium aluminum silicate,
K: potassium aluminum silicate

4. Conclusions

From the strength results, the optimum alkali activator, CP, and temperature of heat treatment for the synthesis of HA-based geopolymer from FA were 0.2 mole of Na_2O/SiO_2 or 0.15 mole of K_2O/SiO_2 with the addition of 2.5 wt.% of CP and treatment at temperature of 500°C. XRD results reveals the appearance of crystalline phase of HA from samples with and without heat treatment. Leachability of Al, Si, Ca, and P from the HA-based geopolymer samples will be investigated for the future work.

Acknowledgement

The National Research Council of Thailand financed this study as part of a larger project to improve the technical and biological features of cement and geopolymer materials for use as biomaterials. (contract number วช.อว.(อ)(กอ)/482/2563). The authors thank the School of Energy, Environment, and Materials and Innovative Environmental Management and Smart Construction Material Laboratory, King Mongkut's University of Technology Thonburi support instrument for analysed.

References

[1] Feng T, Zhu Y, Kozloff K M, Khoury B, Xie Y, Wang X, Cao M, Yuan J, Ta D and Cheng Q 2020 Applied Sciences **10** 22 8214
[2] Scapin M A, Guilhen S N, Cotrim M E and Pires M A F 2015 INAC **47** 5
[3] Cox S C, Mallick K K and Walton R I 2014 Nanobiomaterials **4** 37–47.
[4] Tippayasam C, Sutikulsombat S, Kamseu E, Rosa R, Thavorniti P, Chindaprasirt P, Leonelli C, Heness G and Chaysuwan D 2019 J Aust Ceram Soc **55** 11–17
[5] Kenneth J D M, Nils R, Mark E S and Alan W 2010 J Mater Sci **45** 999–1007
[6] Hervé K T, Daniel F, Claus H R, Elie K, Jean N Y D, Maria C B and Cristina L 2018 Construction and Building Materials **163** 776-792
[7] ASTM D1633-17
[8] Davidovits J 2008 Geopolymer Chemistry and Applications, Geopolymer Institute 584
[9] Abdullah M M A., Hussin K, Bnhussain M, Ismail K N and Ibrahim W M W 2011 Journal of Asian Scientific Research, Asian Economic and Social Society **1** 5 247-253
[10] Derrien A C, Oudadesse H, Sangleboeuf J-C, Briard P and Lucas-Girot A 2004 Journal of thermal analysis and calorimetry **75** 3 937-946
[11] Oudadesse H, Derrien A C, Martin S, Chaair H and Cathelineau G 2008 Applied Surface Science, **255** 2 593-596

IOP Conf. Series: Materials Science and Engineering 1234 (2022) 012026 doi:10.1088/1757-899X/1234/1/012026

Monte Carlo simulations of nanotube filler in composite material: code optimization

N Kerdkaen[1,2,3], T Sutthibutpong[2,3,4], S Phongphanphanee[2,5,6], S Boonchui[1,6], and J Wong-ekkabut[1,2,3,6,*]

[1] Department of Physics, Faculty of Science, Kasetsart University, Bangkok 10900, Thailand.
[2] Computational Biomodelling Laboratory for Agricultural Science and Technology (CBLAST), Faculty of Science, Kasetsart University, Bangkok 10900, Thailand
[3] Thailand Center of Excellence in Physics (ThEP Center), Ministry of Higher Education, Science, Research and Innovation, Bangkok 10400, Thailand
[4] Department of Physics, Faculty of Science, King Mongkut's University of Technology Thonburi (KMUTT), Bangkok 10140, Thailand
[5] Department of Material Science, Faculty of Science, Kasetsart University, Bangkok 10900, Thailand
[6] Specialized Center of Rubber and Polymer Materials for Agriculture and Industry (RPM), Faculty of Science, Kasetsart University, Bangkok 10900, Thailand

*Corresponding E-mail: jirasak.w@ku.ac.th

Abstract The electrically conductive polymer composites (CPCs) have attracted intensive attention for several decades due to their flexibility and unique electrical properties. CPCs are potentially used in many applications such as flexible electrodes, batteries, and strain sensors. The percolated conductive pathways are formed by conductive filler in polymer matrix which is a major effect on the electrical behavior of CPCs. Computational simulations have been used to study the percolation phenomena of CPCs. The simulation algorithms need to be developed and optimized for reducing the simulation time-consuming. In this study, the in-house Monte Carlo simulation that used to estimate percolation threshold is optimized. To simulate in the large-scale system, cut-off distance will be defined to avoid unnecessary complex calculations. The calculation sequence within the code has been rearranged to omit the unnecessary calculation processes. Results show that the optimized software takes less processing time than the previous version around 5 times. Therefore, we can perform the large system to investigate the percolation phenomenon with less lattice confinement effect.

1. Introduction

The electrically conductive polymer composites (CPCs) are the composite materials formed by filling the electrically conductive nanoparticles, such as carbonaceous or metallic particles into the insulated polymer matrix [1]. The CPCs have attracted huge scientific attention for several decades due to their flexibility and unique electrical properties. The properties of CPCs were investigated by varying the types of conductive fillers and insulative polymers. However, the experimental studies were limited when observing the behaviours at the microscopic level. Computational simulations have the advantage to study the properties and behavior of CPCs at the microscopic level. Molecular dynamic (MD) simulations were used to investigate the interactions of biological molecules [2-6]. Recently, MD simulations have been extended to examine the relationship between microscopic behaviours and the macroscopic properties of composite materials [7, 8]. Moreover, the free energy of the nanofiller aggregation in polymer matrix was calculated as a function of the nanofiller concentrations.[8] Nanofiller aggregation plays an important role in the electrical percolation behaviour of CPCs, however the percolation behaviour is still difficult to study by using MD simulations because of time-consuming. In addition, Monte Carlo (MC) technique is more effective to simulate the random dispersion of nanofiller in polymer matrix and to determine the percolation behaviour of CPCs. [9]. Therefore, we began to investigate percolation behaviour in CPCs by using in-house Monte Carlo simulation [10]. The results show that percolation probability depends on filler concentration and the percolation threshold slightly depends on system size due to the confinement effect. In order to compare the simulation results with experiments, the large-scale system of CPCs is needed. Therefore, the simulation program needs to be optimized to reduce computing time. In this study, the in-house Monte Carlo simulation has been redeveloped to estimate percolation threshold. Cut-off distance is applied, the calculation sequence has been rearranged, and the symmetry record of contacted fillers has been used. We found that the MC simulation with 3 optimized techniques could speed up the program.

2. Methodology

MC simulation of nanotubes

The in-house Monte Carlo simulation of nanotubes consists of three main steps, generating the nanotubes into system, checking the contact of all nanotubes, and searching for the percolation paths. The program starts with filling the nanotubes into system, the distances between all nanotubes are calculated to ensure that no overlapping between nanotubes in the system. Next, the program checks the contact between all nanotubes in system by calculating the distance between them. The nanotubes are considered as "in contact" if they are closer than the tunnelling distance of one nanometer. If the contacts are detected, the pair of nanotubes will be recorded. Finally, the percolation paths in system were searched for based on the connected contact of nanotubes. These processes are repeated over and over in which the most significant computing time is the calculations of the distances between nanotubes. Therefore, our improvement is focused on using the distance cut-off, rearranging calculations sequence, and recording the symmetry of contact fillers to remove the unnecessary calculations and speed up the simulations.

Cut-off distance

Calculating the distance between nanotubes is the most complex calculation in program [11]. This calculation sequence is important to identify the contact between nanotubes. However, it costs a huge computing time to calculate the distances between each pair of nanotubes in the system. In optimized version, the distance between the centers of nanotubes is calculated and defined as cut-off distance. If the distance between the center of nanotubes is longer than the length of nanotube, implying that the nanotubes are not in contact without needed to have the additional complex calculations. But if the distance between centers is shorter than the nanotube length, the nanotubes may be in contact and the additional complex calculations are needed to check.

Rearrange the calculation sequence

In our simulation, there are three possible contacts patterns, side-to-side, side-to-end, and end-to-end (As shown in Figure 1) [11]. Each of them has its specific calculation method. Every pair of nanotubes in the system needed to be calculated through these methods. However, the simulation results in Figure 2 show that three quarters of the contact pattern in the system is side-to-side. Therefore, we rearrange the calculation sequence in the program. The optimized version program calculates and checks on side-to-side contact first. If any pair is confirmed that they have a side-to-side contact, the side-to-end and end-to-end contact calculation sequence will be skipped.

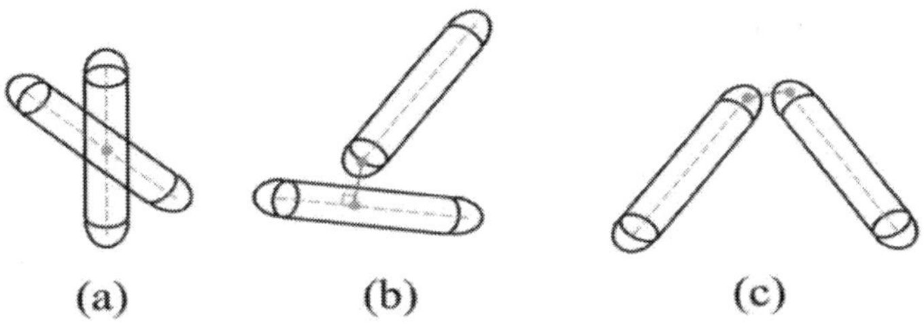

Figure 1. Three possible contact patterns for nanotube consist of (a) side-to-side, (b) side-to-end and (c) end-to-end.

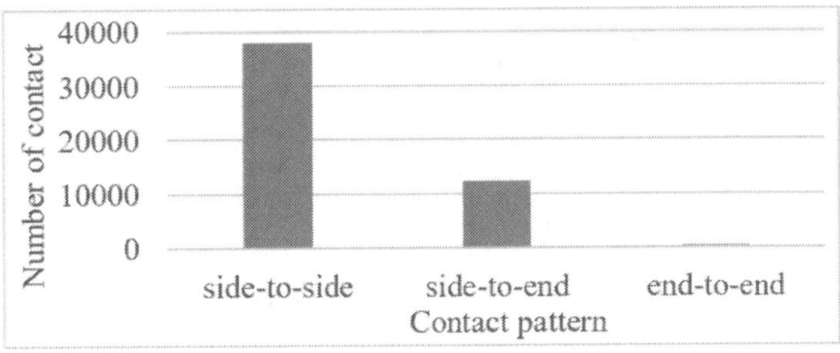

Figure 2. Number of contact for each contact pattern. Data collect from 100 example systems. 75% is side-to-side contact pattern, 24% is side-to-end contact pattern, and 1% is end-to-end contact pattern.

Symmetry record of contacted fillers
In previous version of our simulation program, the contact of all nanotubes in the system was checked via the distance calculation. However, if nanotube A contacts with nanotube B, then nanotube B is also in contact with nanotube A. Therefore, the program does not need to calculate the contact of nanotube B with nanotube A. In optimized version, the program uses the distance calculation to check on a half of contacts and use symmetry record to check on another half.

For example, there are 6 nanotubes in an example system with a label A to F for each nanotube. In previous version, the program use the distance calculation method to check the contact between all nanotubes (A with A to F, B with A to F, C with A to F, …). In optimized version, the program use only checks the contact between the nanotube and following nanotubes (A with B to F, B with C to F, C with D to F, …). The remaining contact is checked via Symmetry record. Such as, if the calculation method confirms that C contact with D, Symmetry record is also records that D contact with C.

3. Results and discussion

The calculation method in the optimized version will remain the same as the previous version. Therefore, the simulation results remain unchanged. For this study, we focus on the time-consuming of the simulation program. The computing time per sample is shown in Figure 3. The largest system with the highest filler concentration in [11] has 2546 nanotubes in a system. Results show that the previous version program takes around an hour and a half to simulate a system with 2500 nanotubes. While it takes less than a quarter of an hour via optimized version. The optimized version of the program takes approximately 5 times less processing time.

In order to simulate the large-scale system of CPCs to estimate percolation threshold without the effect of confinement system size. We target to simulate the system with more than 50,000 nanotubes as the large-scale system of CPCs. Prediction computing time in Figure 3 shows that the previous version program would take more than 700 h to simulate a large-system of CPCs. That time-consuming is unacceptable. However, the optimized version program can simulate a large-scale system of CPCs within around 120 h. Therefore, we are able to simulate the large-scale system of CPCs with acceptable time-consuming.

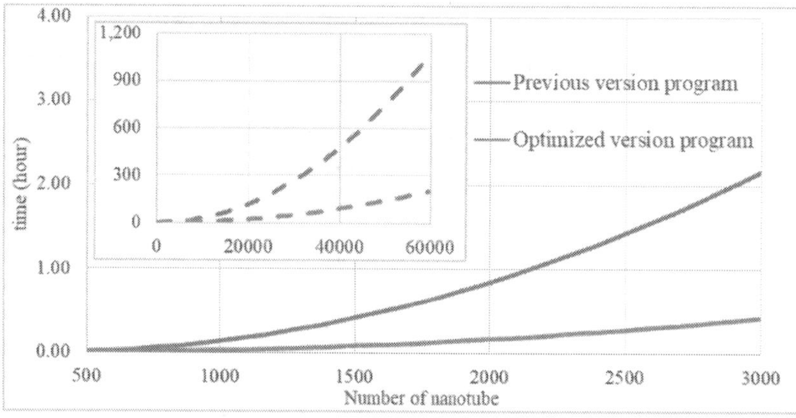

Figure 3. The computing time per sample. Solid lines present the results from program testing and dotted lines present the prediction computing time at the high number of nanotubes.

4. Conclusions

The in-house Monte Carlo simulation of nanotube is successfully optimized to calculate the percolation threshold for the large system. The optimized version program has been improved and has better performance. The results of computing time per sample showed that the optimized version can be simulated the large-system of CPCs with acceptable time-consuming. Therefore, we can perform the large-system simulation to investigate the percolation phenomenon with less confinement effect, which can compare with experiments results. Our in-house program can be applied to other filler type such as filler particle.

Acknowledgement

This work was supported by the National Research Council of Thailand (NRCT) and Specialized Center of Rubber and Polymer Materials in Agriculture and Industry (RPM), Faculty of Science, Kasetsart University. Computing facilities were provided by the Computational Biomodelling Laboratory for Agricultural Science and Technology (CBLAST), Department of Physics, Faculty of Science, Kasetsart University.

References

[1]. Chen J, Zhu Y, Huang J, Zhang J, Pan D, Zhou J, Ryu E. J, Umar A and Guo Z 2021 *Polymer Reviews,* **61** 157.
[2]. Boonnoy P, Karttunen M and Wong-Ekkabut J 2018 *Journal of Physical Chemistry B,* 2018 **122** 10362.
[3]. Wong-ekkabut J and Karttunen M 2016 *Journal of Biological Physics,* **42** 133.
[4]. Nalakarn P, Boonnoy P, Nisoh N, Karttunen M and Wong-ekkabut J 2019 *Scientific Reports,* **9** 1037.
[5]. Nisoh N, Jarerattanachat V, Karttunen M and Wong-ekkabut J 2020 *Biochimica et Biophysica Acta – Biomembranes,* **1862** 183328.

[6]. Khuntawee W, Amornloetwattana R, Vongsangnak W, Karttunen M and Wong-Ekkabut J 2021 *RSC Advances,* **11** 8475.
[7]. Khuntawee W, Sutthibutpong T, Phongphanphanee S, Karttunen M and Wong-Ekkabut J 2019 *Physical Chemistry Chemical Physics,* **21** 19403.
[8]. Kitjanon J, Khuntawee W, Phongphanphanee S, Sutthibutpong T, Chattham N, Karttunen M and Wong-Ekkabut J 2021 *Polymers,* **13** 4044.
[9]. Wenxiang X, Xianglong S and Yang J 2016 *Phys. Rev. E,* **94** 032122.
[10]. Kerdkaen N, Sutthibutpong T, Phongphanphanee S, Boonchui S and Wong-ekkabut J 2019 *IOP Conf. Ser.: Mater. Sci. Eng,* **526** 012006.
[11]. Belashi A 2011 *Percolation modeling in polymer nanocomposites*

Properties of hydroxyapatite-based geopolymer synthesized from calcined kaolin

K Poolkwan[1]*, S Asavapisit [1]* and R Piyapanuwat[2]

[1] Environment Technology Program, School of Energy, Environment and materials, King Mongkut's University of Technology Thonburi, Bangkok, Thailand.
[2] Innovative Environmental Management and Smart Construction Material Laboratory, King Mongkut's University of Technology Thonburi (Ratchaburi Learning Park), Rang Bua, Chom Bueng, Ratchaburi, Thailand.

*Corresponding author: kanyapak.p@mail.kmutt.ca.th, suwimol.asa@kmutt.ac.th

Abstract. This research synthesized the geopolymer from calcined kaolin (600°C for 2 h in a muffle furnace) using sodium hydroxide (NaOH) and potassium hydroxide (KOH) as alkali activators. Calcium phosphate (CP) and heat treatment were introduced to study their effects on the unconfined compressive strength (USC) and crystalline phases formation of kaolin-based geopolymer. Results showed that the highest USC strength at 28 days of geopolymer was obtained from 0.4 moles of Na_2O/SiO_2 and K_2O/SiO_2 and gave strength of 12.95 and 11.56 MPa, respectively. The USC of geopolymer was observed to decrease with the addition of CP. The amounts of 5 wt.% of CP gave a 28-day strength of kaolin-based geopolymer containing NaOH and KOH activators at 9.95 and 8.88 MPa, respectively. The CP-doped geopolymers were then heated at temperatures of 500, 600, and 700°C for 2 h Results indicated that heat treatment at 600°C gave the highest USC of CP-doped geopolymer containing NaOH and KOH equal to 7.61 and 4.61 MPa, respectively with the appearance of hydroxyapatite, muscovite, and quartz in the XRD diffractograms.

1. Introduction

The fractures of bone could be caused either by minor or serious injuries. Sports injuries, traffic accidents and accidents from those that fell from a height are serious injuries that cause several fractures. Bone injuries from a fracture that cannot self-recover is usually need replacement. Bone replacement is also has an increasing demand for an increase elderly population worldwide suffered from osteoporosis. Geopolymers are inorganic materials which can be synthesized from materials containing silica and alumina. These includes calcined kaolinite clay and halloysite clay or a variety of industrial. These raw materials are activated with highly alkaline solutions such as NaOH and KOH with controlled water contents and molar compositions [1]. The Si/Al ratios of the starting materials also play a key role in geopolymerization and geopolymer properties. The optimum geopolymer properties are obtained with molar ratios $SiO_2/Al_2O_3 \sim 3.3$, $H_2O/Na_2O \sim 10$ and $Na_2O/SiO_2 \sim 0.3$ [2]. The network structure of geopolymers act as support for hydroxyapatite (HAP) which is widely used as biocompatible material. It belongs to calcium phosphate (CP) compound with chemical formula of $Ca_{10}(PO_4)_6(OH)_2$. HAP is a material of varying properties depending on the preparation methods with recommended Ca/P molar ratios ranging between 1.50 and 1.67 [3].

HAP-based geopolymer has received an increase interest to be used as bone replacement material for alumina, zirconia, TZP, Mg-PSZ, TiO_2 and bioglasses due to its biocompatible property [4]. This research is therefore aimed at synthesizing HAP-based geopolymer from kaolinite clay. The appropriate molar ratios of Na_2O/SiO_2 and K_2O/SiO_2 for the synthesis geopolymer, the optimum amounts of calcium phosphate and heating temperatures were investigated. The qualities of HAP-based geopolymer were determined from strength development and microstructural observations.

2. Experimental

2.1 Materials

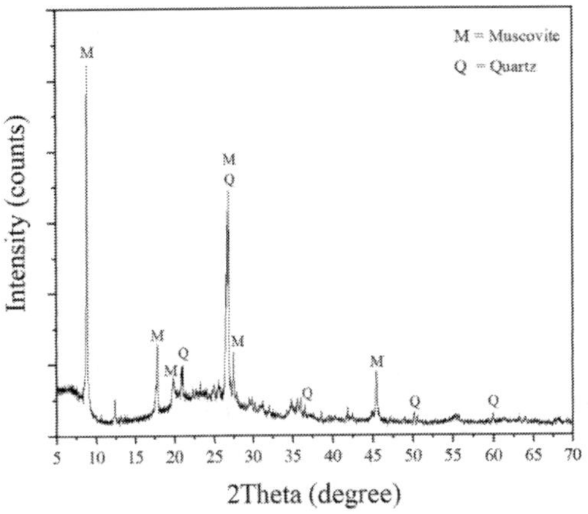

Figure 1. XRD pattern of CK.

Kaolin (K) from the Mineral Resources Development Co., Ltd., Ranong province was used as starting material for the synthesis of geopolymer. The clay was calcined at 600°C for 2 h in a muffle furnace and the CK was ground using ball milling for 2 h. The ground CK was wet sieving until it was retained on sieve No. 325 less than 34% following ASTM C618-12a. The chemical compositions of CK were analyzed using X-ray Fluorescence Spectrometer (XRF) technique and is shown in Table 1. Oxides of silica (SiO_2) and alumina (Al_2O_3) are the two main elements present in the ground CK at 53.1 and 40.8 wt.%, respectively. XRD pattern shows muscovite ($KAl_2(Si,Al)_4O_{10}(OH)_2$) and quartz ($SiO_2$). Commercial grade chemicals in this research, including NaOH from Formosoda-p, Taiwan, KOH from UNID, Korea and CP ($Ca_3(PO_4)_2$) from FDSCL, Mumbai having purity of 99, 95 and 99%, respectively were used.

Table 1. Chemical compositions of ground CK.

	Chemical compositions (%)							
	SiO_2	Al_2O_3	K_2O	Fe_2O_3	MgO	TiO_2	CaO	LOI
CK	53.1	40.8	3.4	2.09	0.1	0.07	0.03	1.25

2.2 Sample preparation

Geopolymer samples were prepared by mixing ground CK with initial Si/Al ratio of 1.15 with Na_2O/SiO_2 and K_2O/SiO_2 at 0.15, 0.2, 0.25, 0.3, 0.35, and 0.4 molar ratios. The mixtures were mixed and transferred into cylindrical PVC molds. Geopolymer samples with NaOH or KOH activators having maximum UCS were selected for further experiment. CP were added to the selected geopolymer samples at 2.5, 5, 7.5, and 10 wt.%. The optimum amount of CP was determined from sample with highest strength and selected for further heat treatment at 500, 600, and 700°C for 2 h.

2.3 Method and testing

XRF (BRUKER AXS, S4PIONEER) were used to analyse the chemical compositions of ground CK. The unconfined compaction machine with a load cell of 5,000 N and handheld data logger (TK-32K) were used to record data following ASTM D1633-17. X-ray Diffraction diffractogram (XRD) (BRUKER AXS, D8DISCOVER) was used for microstructure observation on the formation of crystalline phases.

3. Results and discussion

3.1 Effect of the alkaline activators on the unconfined compressive strength of calcined kaolin based geopolymer

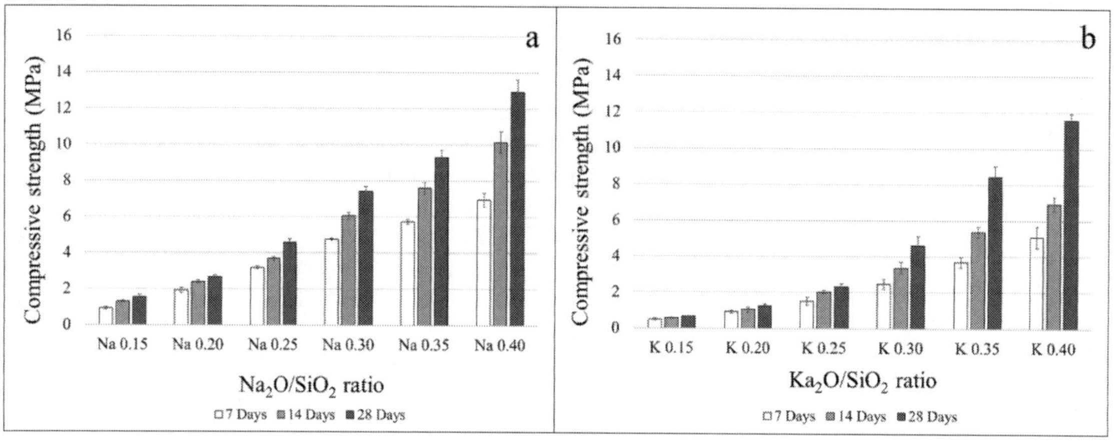

Figure 2. Effect of alkaline activators on UCS of CK based geopolymer (a) NaOH (b) KOH.

Variation of NaOH solution at Na_2O/SiO_2 molar ratio on UCS development of CK-based geopolymer is shown in Figure 2(a). At 7 days of curing, UCS at Na_2O/SiO_2 molar ratio of 0.15, 0.2, 0.25, 0.3, 0.35, and 0.4 was 0.93, 1.91, 3.20, 4.77, 5.74 and 6.95 MPa, respectively and increased to 1.54, 2.66, 4.58, 7.43, 9.28 and 12.95 MPa after curing for 28 days. Figure 2(b) shows the strength development of CK-based geopolymer with KOH at K_2O/SiO_2 molar ratio. Strength of CK-based geopolymer after curing for 7 days was 0.49, 0.92, 1.50, 2.45, 3.69 and 5.08 MPa. and increased to 0.66, 1.23, 2.30, 4.62, 8.42 and 11.56 MPa at the age of 28 days for samples containing K_2O/SiO_2 at molar ratio of 0.15, 0.2, 0.25, 0.3, 0.35 and 0.4 mol, respectively. The maximum strength was obtained from CK-based geopolymer with Na_2O/SiO_2 and K_2O/SiO_2 molar ratio of 0.4 and gave strength of 12.95 and 11.56 MPa after curing for 28 days. Results showed that strength increased with the increase of Na_2O/SiO_2 or K_2O/SiO_2 molar ratio and curing time. The strength of all samples containing KOH were lower than NaOH.

Strength development of CK-based geopolymer is resulting from the high alkalinity of NaOH and KOH activators leading to the dissolution of SiO_2 and Al_2O_3 from CK. The soluble silicates and aluminates react with Na^+ and K^+ cations from the alkali activator and produced sodium or potassium aluminosilicate gel [5] which contributes to strength. In addition, a smaller ionic radius of Na^+ cation compares to K^+ cation play an essential role in an increase solubility from CK. As a result, an increase degree of geopolymerization was obtained, and thus higher strength of CK-based geopolymer activated with NaOH was observed. From the experimental results, CK-based geopolymers with Na_2O/SiO_2 and K_2O/SiO_2 molar ratio of 0.4 were selected for further study.

3.2 Effect of calcium phosphate on the compressive strength of hydroxyapatite based geopolymer

Strength development of CK-based geopolymer activated by NaOH as a function of various amounts of CP is shown in Figure 3(a). The addition of CP at all levels resulted in a reduction of strength development of CK-based geopolymer at all curing duration. A similar observation was reported by previous research and found that the inclusion of CP increased porosity during geopolymerization process leading to a decrease of strength [6,7] The highest strength was obtained at 5 wt.% of $Ca_3(PO_4)$ addition and gave strength 9.95 MPa at 28 days. Similar variations in strength development were obtained from CK-based geopolymer activated by KOH when $Ca_3(PO_4)_2$ was added at 0, 2.5, 5.0, 7.5

and 10.0 wt.% shown in Figure 3(b). Results showed that strength development of geopolymer activated by KOH was lower than NaOH at all levels of $Ca_3(PO_4)_2$ addition and at all curing durations. Tchakouté *et al.* [8] reported that CK-based geopolymer mixed with 0, 2, 4, 6, 8 and 10 wt.% of $CaHPO_4$ gave the highest USC 48.4 MPa at 6 wt.% addition of $CaHPO_4$. The reduction of the USC related to the excess PO_4 units in the system. This phenomenon results in an imbalanced charge of $(PO_4)^{3-}$ ions leading to a weakening of the microstructure and thus reduce strength of the geopolymers. Therefore, the addition of 5 wt.% of $Ca_3(PO_4)_2$ to CK-based geopolymer activated with NaOH and KOH was selected for the heat treatment study.

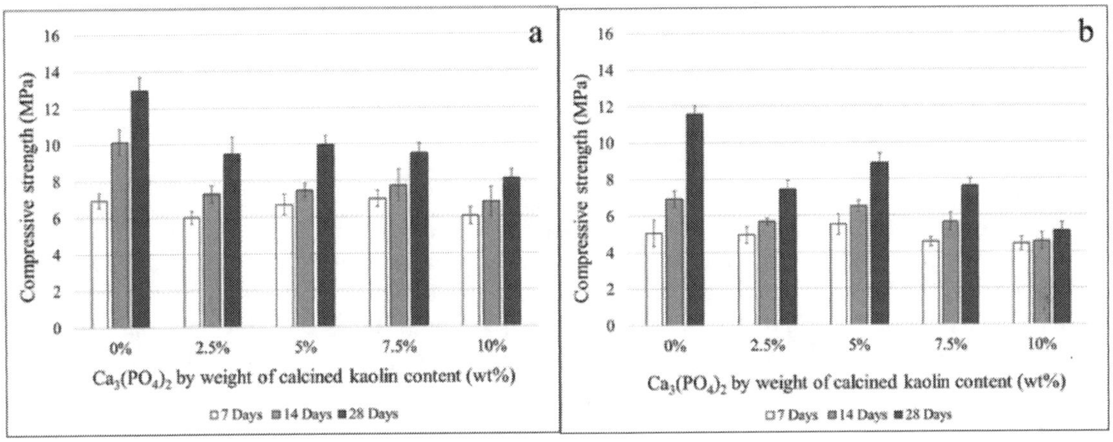

Figure 3. UCS of CK-based geopolymer with $Ca_3(PO_4)_2$ (a) NaOH activator (b) KOH activator.

3.3 Effect of heat treatment on strength of CK-based geopolymer

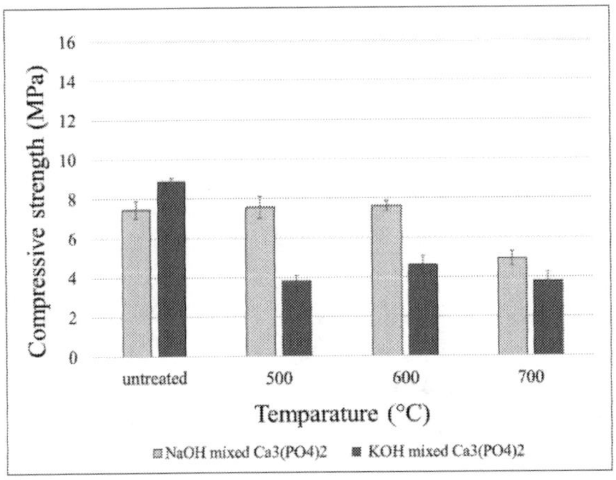

Figure 4. UCS of CK-based geopolymers after heat treatment under various temperatures.

CK-based geopolymer activated added 5 wt.% of $Ca_3(PO_4)_2$ with NaOH and KOH gave the strength of 7.51, 7.61, 4.92 and 3.78, 4.61, 3.77 MPa when subjected to heat treatment at 500, 600 and 700°C, respectively. (Figure 4) The highest strength was found at 600°C for CK-based geopolymers activated with both NaOH and KOH. Thokchom *et al.* [9] investigates the effects of heat treatment between 300 to 700°C on the pore structures of geopolymer paste. In addition, Kenneth *et al.* [2] reported that heat treatment reduced cracking of geopolymer and pH value of simulated body fluid (SBF) after soaking.

3.3 XRD results of CK-based geopolymer

The XRD patterns of CK-based geopolymer with both NaOH and KOH and addition of 5 wt.% of $Ca_3(PO_4)_2$ (CKN and CKK) were shown in Figures 5. Peaks of sodium aluminum silicate (SA), HAP, muscovite and quartz appeared in CKN. The formation of SA resulted in development of UCS of CK-based geopolymer. The appearance of HAP, potassium aluminum silicate (PA), calcium phosphate hydrate, muscovite, and quartz were observed from CKK. The SA and PA developed UCS of CKN and CKK. XRD pattern of CK-based geopolymer with both NaOH and KOH and added 5 wt.% of $Ca_3(PO_4)_2$ after heat treatment at temperature of 600°C (CKN600 and CKK600). Heat treatment reduced the

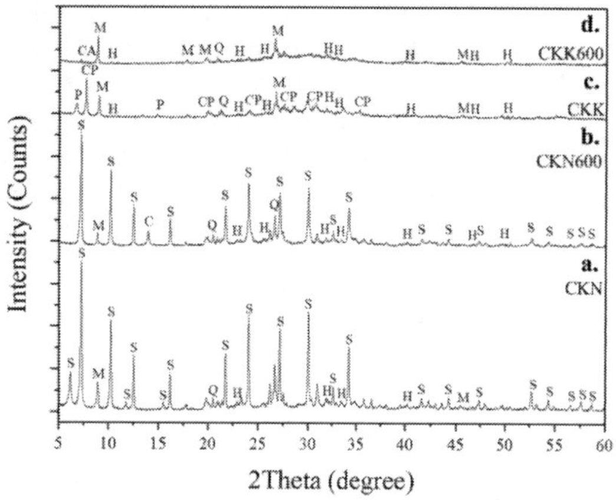

intensity of the peaks for SA, HAP, muscovite, and quartz with the appearance of calcium aluminum silicate hydrate (CASH) in CKN600. While XRD pattern of CKK600 found HAP, calcium aluminum silicate, muscovite and quartz. Other research found HAP after soaking in SBF solution [2, 10].

Figure 5. XRD pattern of CK-based geopolymer with 5 wt.% $Ca_3(PO_4)_2$ addition (a) NaOH activator (b) NaOH activator after heat treatment at 600°C (c) KOH activator (d) KOH activator after heat treatment at 600 °C

4. Conclusions

This research found that the highest strength of CK-based geopolymer was obtained from a sample with Na_2O/SiO_2 and K_2O/SiO_2 molar ratio of 0.4, at 5 wt.% addition of CP and heat treatment at the temperature of 600°C. Heat treatment at the temperature of 600°C resulted in the formation of crystalline hydroxyapatite within the geopolymer network synthesized from ground CK which play an important role as support for hydroxyapatite. Further research is on-going to study the suitability of using hydroxyapatite-based geopolymer as bio-ceramic materials to replace bone.

Acknowledgement

This research was supported by the Research Fund of the National Research Council of Thailand (contract number วช.อว.(อ)(กอ)/482/2563).

References

[1] Duxson P, Ferna´ndez-Jime´nez A, Provis J L, Lukey G C, Palomo A and Deventer van J-S-J 2007 *J Mater Sci* 42 2917-2933
[2] Kenneth J D M, Nils R, Mark E S and Alan W 2010 *J Mater Sci* **45** 999–1007
[3] Fihri A, Len C, Varma R S and Solhy A 2017 *Coordination Chemistry Reviews* **347** 48-76
[4] Ebadzadeh T, Behnamghader A and Nemati R 2011 *Ceramics International* **37** 2887-2889
[5] Komnitsas K, Zaharaki D 2007 *Miner. Eng.* **20** 1261-1277
[6] Murri A N, Medri V, Papa E, Laghi L, Mingazzini C and Landi E 2017 *Environments* **4** 86
[7] Subaer 2004 PhD *Thesis* Curtin University of technology
[8] Tchakouté H K, Fotio D, Rüscher C H, Kamseu E, Djobo J N Y, Bignozzi M C and Leonelli C 2018 *Construction and Building Materials* **163** 776-792
[9] Thokchom S, Mandal K K and Ghosh S 2012 *Arab J Sci Eng* **37** 977-989
[10] Pangdaeng S, Sata V, Aguiar J B, Pacheco T F and Chindaprasirt P 2015 *Materials Science and Engineering C* **51** 1–6

Electrochemical performance of binder-free and flexible spinel $NiCo_2O_4$ electrode

V Somsongkul[1,2*], A Thong-em[1], P Chaojeen[1], and C Kongmark[3,*]

[1] Department of Industrial Chemistry, Faculty of Applied Science, King Mongkut's University of Technology North Bangkok, Bangkok 10800, Thailand.
[2] Integrated Chemistry Research Center for Sustainable Technology (ICRT), King Mongkut's University of Technology North Bangkok, Bangkok 10800, Thailand
[3] Specialized Center of Rubber and Polymer Materials in Agriculture and Industry (RPM), Department of Materials Science, Faculty of Science, Kasetsart University, Bangkok 10900, Thailand.

*Corresponding author: voranuch.s@sci.kmutnb.ac.th, chanapa.k@ku.th

Abstract. This research aims to study the one-step hydrothermal synthesis of a binder-free and flexible spinel nickel cobaltite ($NiCo_2O_4$) electrode for supercapacitor application. The hydrothermal synthesis method is well known to generate highly crystalline products, with high purity, narrow size distribution, and low aggregation of particles. The spinel $NiCo_2O_4$ nanoparticles were grown on a flexible carbon cloth substrate under hydrothermal conditions at 130°C for various reaction times (1, 3 and 6 h). The morphology of $NiCo_2O_4$ nanoparticles was tuned from nanosheet to nanowire by varying the hydrothermal reaction time from 1 h to 6 h. The electrochemical properties of $NiCo_2O_4$ electrodes were investigated. The $NiCo_2O_4$ electrode prepared for 6 h exhibited the highest anodic and cathodic current density and the lowest charge transfer resistance compared with the other electrodes.

1. Introduction

Supercapacitors are electrochemical energy storage devices beyond batteries because of their low internal resistance, a very high capacity, and a fast charge-discharge rate [1, 2] which provide superior properties including their high power, wide thermal range (-40 to 70°C), long life, low maintenance, flexible packaging, and low weight [3]. These devices are fully integrated in the essential application areas of a short load cycle and high reliability requirement such as portable electronic devices and environmental friendly energy storage such as fuel cell vehicles and low emission hybrid vehicles. A supercapacitor consists of two electrodes (anode and cathode), an electrolyte, and a separator which electrically isolates the two electrodes. The mechanisms of supercapacitors can be classified into three categories: (i) electrical double layer capacitor (Helmholtz) includes physical adsorption/desorption on the electrolyte/electrode interface based on electrostatic charge diffusion and accumulation at the electrode/electrolyte interface, which strongly relies on electrode surface area that is available for electrolyte ions. (ii) pseudocapacitor exhibits a fast and reversible faradic process due to the presence of electroactive species (redox reaction). (iii) hybrid combines the properties of both electrical double layer capacitor and pseudocapacitors [3]. The electrode materials play an important on the performance of supercapacitors. The most common used materials are carbon materials, transition metal oxides and

IOP Conf. Series: Materials Science and Engineering 1234 (2022) 012028 doi:10.1088/1757-899X/1234/1/012028

conducting materials. Transition metal oxides have attracted considerable attention in the field of energy storage/conversion not only because of their beneficial reported mechanical, structural, or electronic properties but also because of their high pseudocapacitance ascribed to their multiple valence state change, which is generally not possible via carbon materials [4]. This work is dedicated to the investigation of spinel nickel cobaltite ($NiCo_2O_4$) nanomaterial for the supercapacitor applications. The important structure feature of the spinel nickel cobaltite is the hybridization of metal valence orbitals, which would promote redox reactions, making it an ideal electrode material for supercapacitor. $NiCo_2O_4$ exhibits high electrical conductivity as well as the electrochemical activity and storage performances [5]. The preparation methods of $NiCo_2O_4$ electrode material strongly influence on the microstructure and physico-chemical properties of the active material. Hydrothermal method is advantageous to generate highly crystalline products, with high purity, narrow size distribution, and low aggregation of nanoparticles [4]. In this work, spinel nickel cobaltite electrode was prepared by the one-step hydrothermal with a binder-free on flexible carbon cloth. The influence of hydrothermal reaction time on morphology and electrochemical performances of spinel $NiCo_2O_4$ electrodes have been investigated.

2. Experimental

The $NiCo_2O_4$ was synthesized on carbon cloth by hydrothermal method. The carbon cloth (CeTech Co., Ltd.) was cleaned by sonicating with acetone, deionized water, and ethanol, respectively, 15 min each and then electrochemically pretreated in 3M HCl aqueous solution. The precursor of $NiCo_2O_4$ was prepared by stirring $Co(NO_3)_2{\cdot}6H_2O$ (99%, Qrec), $Ni(NO_3)_2{\cdot}6H_2O$ (98%, Alfa Aesar), NH_4F (98%, Sigma-Aldrich), and urea (98%, Alfa Aesar) in deionized water for 40 min until a homogeneous solution was obtained. Then, the mixed solution and the cleaned carbon cloth were transferred into autoclave. The hydrothermal reaction was carried out at 130°C for various reaction times (1, 3 and 6 h). Afterwards, the precursor on carbon cloth was rinsed with deionized water and ethanol and sintered in the air at 400°C for 4 h. The flexible $NiCo_2O_4$/carbon cloth electrode obtained after 1 h, 3 h and 6 h reaction times are named as 1h-$NiCo_2O_4$, 3h-$NiCo_2O_4$ and 6h-$NiCo_2O_4$, respectively. The structure of materials was studied by X-ray diffraction analysis (Bruker D8 Advance). The morphology of materials was investigated by scanning electron microscopy (FEI Quanta 450). The electrochemical analyses (cyclic voltammetry and electrochemical impedance spectroscopy) were performed by using Zahner electrochemical workstation IM6. The specific capacitance was calculated according to the equation [6]:

$$C_s = \frac{\int idV}{2ms\Delta V}$$

where C_s is the specific capacitance, $\int idV$ is the integrated area under a curve, s is the potential scan rate in V/s, ΔV is the potential range and m is the mass of the active material electrode.

3. Results and discussion

Figure 1. presents XRD spectra of $NiCo_2O_4$ grown on carbon cloth at various reaction time (1, 3 and 6 h). All samples show diffraction peaks at 18.92, 31.15, 36.67, 38.40, 44.67, 59.09, and 64.93 which correspond respectively to the (111), (220), (311), (222), (400), (422), (511), and (440) planes of the spinel $NiCo_2O_4$ structure with Fd-3m space group (JCPDS No. 01-073-1702). Moreover, the average crystallite sizes of the 1 h, 3 h, and 6 h $NiCo_2O_4$ particles, estimated from the XRD peak width by using the Debye Scherrer equation, are 9.5 ± 0.7, 9.5 ± 0.3 and 8.9 ± 0.2 nm, respectively. The morphology of $NiCo_2O_4$ nanostructures on carbon cloth substrates changes with hydrothermal reaction time. It can be seen from micrographs at 2,500X magnification (Figure 2.) that the formation of the $NiCo_2O_4$ nanoparticles increases with increasing reaction time. After 1 h of reaction time, a few nanosheets grew on the surface of the carbon cloth. Two hours later, a greater amount of nanosheets was observed, accompanied with nanowires. After 6 h of reaction, the carbon cloth is fully covered with $NiCo_2O_4$ nanowires, which would provide a good interfacial contact between conductive substrate and active material, resulting in high conductivity.

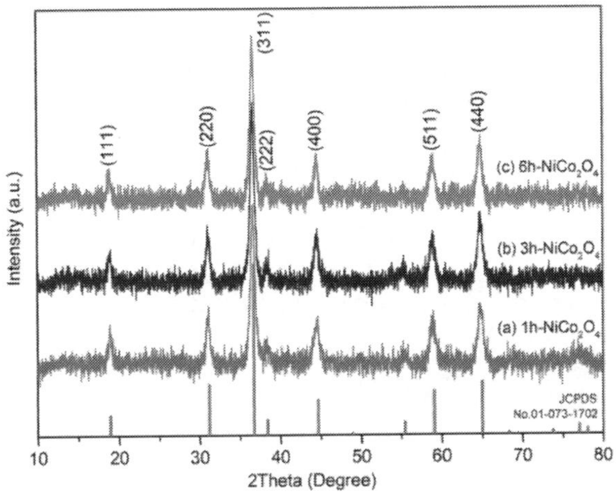

Figure 1. XRD spectra of $NiCo_2O_4$ prepared by a hydrothermal method at various reaction times (a) 1 h, (b) 3 h, and (c) 6 h.

Figure 2. SEM images of $NiCo_2O_4$ nanoparticles prepared by a hydrothermal method at various reaction times (a) 1 h, (b) 3 h, and (c) 6 h.

The electrochemical properties of the $NiCo_2O_4$ were investigated as working electrode in a three-electrodes system with an aqueous 6 M KOH electrolyte as shown in Figure 3. All $NiCo_2O_4$ electrodes exhibit redox peaks, as shown in Figure 3(a), which indicates pseudocapacitive characteristic of the supercapacitor. There are no redox peaks for the carbon cloth without active materials. The charge storage mechanism of $NiCo_2O_4$ during redox reaction corresponds to the reversible reactions of Co^{2+}/Co^{3+} and Ni^{2+}/Ni^{3+} transitions as given in the following equations [5]

$$NiCo_2O_4 + OH^- + H_2O \rightleftharpoons NiOOH + 2CoOOH + e^- \quad (1)$$
$$MOOH + OH^- \rightleftharpoons MO_2 + H_2O + e^- \quad (2)$$

where M is transition metal of Co or Ni. The CV integrated area of anodic and cathodic peak increases with increasing reaction time. It seems likely that the more the active material is formed and deposited on carbon cloth, the better conductivity and electron transfer at the electrode/electrolyte interface the electrode achieves. The mass of active material on the 1 h, 3 h and 6 h $NiCo_2O_4$ electrode are 0.8, 0.9 and 2.2 mg, respectively. The calculated specific capacitances of the 1 h, 3 h, and 6 h $NiCo_2O_4$ electrode are 38.63, 73.79, and 36.48 F g^{-1}, respectively. It is possible that the growth of $NiCo_2O_4$ nanowires in the 6 h $NiCo_2O_4$ electrode may be too thick or dense, therefore not all particles involve in the electrochemical reaction. Figure 3(b), the shape of their CV curves maintains well at various scan rates and the redox peak currents rise with increasing scan rates, which would suggest the fast faradaic reaction kinetics on the electrode surface. Meanwhile, the anodic and cathodic peaks shift toward negative and positive potentials, respectively. The EIS analysis of the $NiCo_2O_4$ electrode was carried out using an equivalent circuit as shown in the Figure 3(c) and (d). All $NiCo_2O_4$ electrodes provide small values of solution resistance (R_s), about 3.0 - 3.3 Ω, indicating low bulk resistance of the system. The fitted semi-circle diameters of $NiCo_2O_4$ electrodes decrease from 1.41 to 0.626 Ω with increasing reaction time from 1 to 6 h. The 6h-$NiCo_2O_4$ electrode has the lowest faradaic charge transfer resistance (R_{ct}) of the electrode/electrolyte interface, indicating the promoted charge transport kinetics in the bulk which is also associated with the highest anodic and cathodic peak currents observed in CV curves. Moreover, the 6h-$NiCo_2O_4$ electrode exhibits the shortest Warburg line and lowest diffusion resistance in the material, which implies a superior diffusion of hydroxide ions into the pores of electrode material, thus improving the capacitive performance [7].

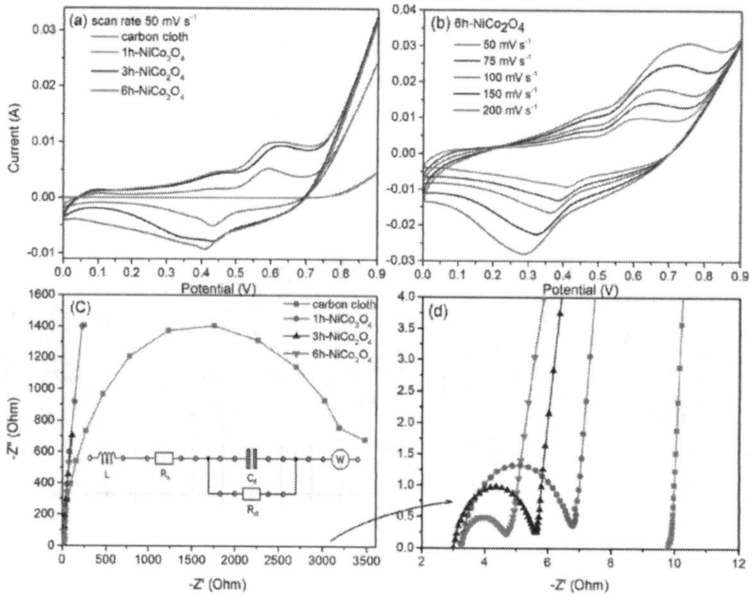

Figure 3. $NiCo_2O_4$ electrode prepared by hydrothermal method (a) CV curves of $NiCo_2O_4$ electrodes collected at various reaction times (b) CV curves of the 6h-$NiCo_2O_4$ electrode at various scan rates, (c) and (d) Nyquist plots of all $NiCo_2O_4$ electrodes.

4. Conclusions

This study has shown that the hydrothermal synthesis method not only provides $NiCo_2O_4$ nanoparticles grown on flexible carbon cloth without any binder but also simplify the electrode preparation process. The $NiCo_2O_4$ electrode prepared with the hydrothermal condition at 130°C for 6 h was found to outperform the electrochemical properties, including the highest anodic and cathodic currents, the highest specific capacitance, the lowest charge transfer resistance and the lowest diffusion resistance compared with other electrodes. The $NiCo_2O_4$ pseudocapacitor electrodes have redox couple, which achieve a rapid transport of ions and electrons and reversible faradic process.

Acknowledgement

This work was supported by the Faculty of Applied Science, King Mongkut's University of Technology North Bangkok. Contract no. 6241102.

References

[1] Kamal K K 2020 Springer Series in Materials Science (Switzerland: Springer) **302** 1.
[2] Hassan I U, Salim H, Naikoo G A, Awan T, Dar R A, Arshad F and Qurashi A 2021 *J. Saudi*

Chem. Soc. **25** 101228.
[3] Iro Zaharaddeen S, Subramani C and Dash S S 2016 *Int. J. Electrochem Sci.* **11** 10628.
[4] Kenneth I O and Shaowei C 2016 Springer Series in Nanostructure Science and Technology (Switzerland: Springer) **6331** 317.
[5] Dubal D P, Gomez-Romero P, Sankapal B R and Holze R 2015 Nano Energy **11** 377.
[6] Kim B K, Chabot V, and Yu A 2013 *Electrochim. Acta,* **109** 370.
[7] Zhu B, Tang S, Vongehr S, Xie H and Meng X 2016 ACS Appl. Mater. Interfaces. **8** 4762.

IOP Conf. Series: Materials Science and Engineering 1234 (2022) 012029 doi:10.1088/1757-899X/1234/1/012029

Solid-state reaction synthesis and characterization of Mn-doped $LiFePO_4$ cathode material

A M M Kyaw[1,2], G Panomsuwan[1,2] and R Munprom[1,2,*]

[1] Department of Materials Engineering, Faculty of Engineering, Kasetsart University, Bangkok 10900, Thailand

[2] International Collaborative Education Program for Materials Technology, Education, and Research (ICE-Matter), ASEAN University Network/Southeast Asia Engineering Education Development Network (AUN/SEED-Net), Bangkok, Thailand

*Corresponding author : fengrtpm@ku.ac.th

Abstract. Olivine type $LiFePO_4$ has great advantages for Li-ion batteries due to its non-toxicity, high safety, and good cycle life performance. However, its low-rate capability and low energy density make some challenges for this $LiFePO_4$. Several methods like doping with transition metals were used, and Mn ion was used in this work to improve the overall electrochemical properties. $LiMn_xFe_{1-x}PO_4$ is promising cathode material owing to high voltage, structural and chemical stability. However, the electrochemical performance of these materials depends on phases and structures obtained from synthesis. In this work, the effect of solid-state reaction conditions, including calcination temperature and duration, on morphology, structure, and electrochemical properties of $LiMn_xFe_{1-x}PO_4$ cathode materials with the composition of $x = 0.5$ was investigated. The morphology, crystallography and local structure of the synthesized materials were examined by field emission scanning electron microscope (FE-SEM), X-ray diffractometer (XRD) and Fourier-transform infrared spectrometer (FTIR), respectively. The surface area was also determined by the Brunauer-Emmett-Teller (BET) model. The effect of calcination temperature and reaction time upon the morphology, structures of the synthesized cathode materials were studied and discussed. The results could be essential for further development and employment of $LiMn_xFe_{1-x}PO_4$ in Li-ion batteries.

1. Introduction

Olivine-type lithium-transition metal phosphates ($LiMPO_4$) (M = Fe, Mn, Co, Ni, etc…) are taken into account for potential cathode materials due to high theoretical specific capacity, superior safety capacity, environmental friendliness and low cost [1]. Among these $LiMPO_4$, $LiFePO_4$ (theoretical capacity of 170 $mAhg^{-1}$ and operation potential of 3.4 V vs. Li/Li^+) is well known for the aforementioned advantages as well as has been investigated intensively and successfully commercialized [2]. However, this material possesses drawbacks of low electronic conductivity ($<10^{-9}$ Scm^{-1}) and poor Li^+ ion diffusion. In addition, its low working potential (3.4 V vs. Li/Li^+) restricts to use in high power applications, such as electric vehicles

IOP Conf. Series: Materials Science and Engineering 1234 (2022) 012029 doi:10.1088/1757-899X/1234/1/012029

[3-5]. Several methods can be used to tackle the drawbacks such as tailoring morphology, coating conductive carbon on primary particles and doping with some transition metal ions (e.g., Mn^+, Co^+, Ni^+, Mg^{2+}, Cr^{3+}, Ti^{4+}, etc.) [6, 7].

Owning the same structure as $LiFePO_4$, lithium manganese phosphate ($LiMnPO_4$), is also a promising alternative cathode material [8]. Containing Mn element gives benefits in terms of material cost and environmental friendliness. Compared with $LiFePO_4$, it exhibits a higher potential of 4.1 V vs. Li/Li^+, which leads to being more suitable for high power applications [9, 10] Nevertheless, structural change during intercalation and deintercalation of Li ions, Jahn-Teller distortion of unstable Mn^{3+} ion and poor electronic conductivity make this cathode material imperfect. Similar improvement methods can be implemented to enhance the properties of $LiMnPO_4$ [11, 12]. For example, Jo *et al.* reported that the composition of $LiMn_{0.71}Fe_{0.29}PO_4$ cathode material prepared by polyol method showed the high volumetric capacity of 243 and 128 mAh cm^{-3} at 0.1C and 7C, respectively, and excellent cycling performance [13]. Moreover, $LiMn_{0.5}Fe_{0.5}PO_4$ prepared by a solvothermal route exhibited long term cycling stability and good high rate performance up to 1000 cycles [9].

Various synthesis methods, such as sol-gel and solid-state reaction, can be used to prepare $LiMn_xFe_{(1-x)}PO_4$ and the synthesis condition can strongly influence electrochemical performance [6, 12, 14]. By adjusting synthesis parameters, the morphology of the materials can be tailored to nanostructure. Among all techniques, a solid-state reaction is often applied due to simplicity and low cost. Many research works have applied this method to synthesize the cathode materials, but most of the works used long two-step calcination at high temperatures, which is quite energy and time consuming [8, 14]. Only a few works demonstrate a synthesis condition using short calcination time. Therefore, in this work, $LiMn_xFe_{(1-x)}PO_4$ with $x = 0.5$ was synthesized by a solid-state reaction method and, then thermally treated at 450°C and 550°C for a given time (i.e., 2, 6 and 12 h). The particle morphology, structure, local structure, and specific surface area obtained from different calcination times was investigated and discussed. This relationship could emphasize the importance of synthesis condition optimization for performance improvement.

2. Experimental

2.1. Synthesis of $LiMn_{0.5}Fe_{0.5}PO_4$ Materials

$LiMn_{0.5}Fe_{0.5}PO_4$ cathode materials were synthesized by a solid-state reaction method. Manganese (II) carbonate (Aldrich, 99.9%), iron (II) oxalate dihydrate (Aldrich, 99%) and lithium phosphate monobasic (Aldrich, 99%) were used as starting materials and ethanol was used as a dispersant. The stoichiometric ratio of Li: Mn: Fe (1: 0.5: 0.5) and an appropriate amount of ethanol were mixed by jar rolling with yttrium-stabilized zirconia balls as grinding media for 18 h. Then, the mixture was dried overnight at 80°C to evaporate the ethanol. After that, the dried powder was ground. Finally, the materials were calcined at 450°C and 550°C for 2, 6 and 12 h in argon (Ar) atmosphere to obtain the final powder of $LiMn_{0.5}Fe_{0.5}PO_4$.

2.2. Materials Characterization

The crystal structure and phase of the synthesized materials were examined using an X-ray diffractometer (Philips X'Pert) with Cu K_α radiation (λ= 0.154 nm) over the range of 10° to 60° at a step size of 0.02°. The surface morphology and particle size of the particles were analyzed using a field emission scanning electron microscope (JEOL-JSM-7600F). The composition of the $LiMn_{0.5}Fe_{0.5}PO_4$ powder was analyzed by energy dispersive spectroscopy (EDS). The Fourier transform infrared (FT-IR) spectrum was obtained with KBr pellets using Bruker Alpha-E spectrometer. The specific surface areas were measured with Micromeritics 3Flex at 77 K using the Brunauer-Emmett-Teller method.

3. Results and discussion

The X-ray diffraction pattern of the cathode materials for each synthesis condition is shown in Figure 1. For all conditions, some peaks were well indexed with the orthorhombic structure with the space group of Pnmb (62) (ICDD-00-042-0580). Such peaks can be identified as $LiMn_{0.5}Fe_{0.5}PO_4$. Some impurity phases, namely $LiPO_3$, Li_3PO_4 and $Li_4P_2O_7$ were also found. However, the impurity phases decreased when the temperature increased. This is possibly caused by an incomplete reaction at low temperatures and short calcination times. The crystallite sizes of these materials were estimated using the Scherrer's equation and shown in Table 1. The crystallite sizes of all synthesis conditions were in the range of 25-30 nm and increased with the calcination time.

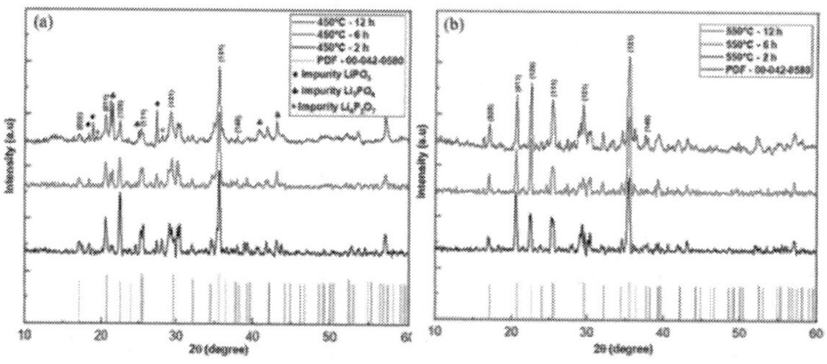

Figure 1. XRD patterns of $LiMn_{0.5}Fe_{0.5}PO_4$ calcined at (a) 450°C (b) 550°C for 2, 6 and 12 h.

The morphology and structure of the synthesized materials were characterized by FE-SEM and the images are shown in Figure 2(a - f). From the results, the particles of all conditions were similar. Specifically, particles were agglomerated with a primary size of 20 – 52 nm. However, the particle sizes became larger with increasing temperatures. On the other hand, increasing calcination time can slightly decrease the particles sizes. The box plot of particle size analyses of $LiMn_{0.5}Fe_{0.5}PO_4$ is shown in Figure 2(g).

Figure 2. FESEM images of $LiMn_{0.5}Fe_{0.5}PO_4$ (a - c) synthesized at 450°C for 2, 6 and 12 h, respectively (d - f) synthesized at 550 °C for 2, 6 and 12 h, respectively and (g) box plot of particle size (yellow and blue color represents calcination at 450°C and 550°C, respectively).

The functional group vibration modes of $LiMn_{0.5}Fe_{0.5}PO_4$ prepared at different conditions were examined by FTIR spectroscopy in the region of 4000 - 400 cm^{-1} and are shown in Figure 3. Both spectra have similar vibrational modes. The bands at 1092 and 1052 cm^{-1} belong to the $LiMnPO_4$ and $LiFePO_4$ respectively. A single broad band found between these two bands belongs to a solid solution of $LiMn_{0.5}Fe_{0.5}PO_4$ [15]. Such a solid solution began to form at high temperatures for a long period of time. As a result, it can only be found in the condition of 550 °C for 12 h, and not found in the other conditions due to the incomplete reaction. In addition, all synthesized materials exhibited the vibrational mode of PO_4^{3-}: three bands between 1140 and 1043 cm^{-1} corresponding to (υ_3) antisymmetric stretching mode, two bands between 978 and 749 cm^{-1} corresponded to (υ_1) symmetric stretching mode, three bands between 636 and 552 cm^{-1} corresponded to (υ_4) bending mode and the band around 460 cm^{-1} corresponding to (υ_2) bending mode.

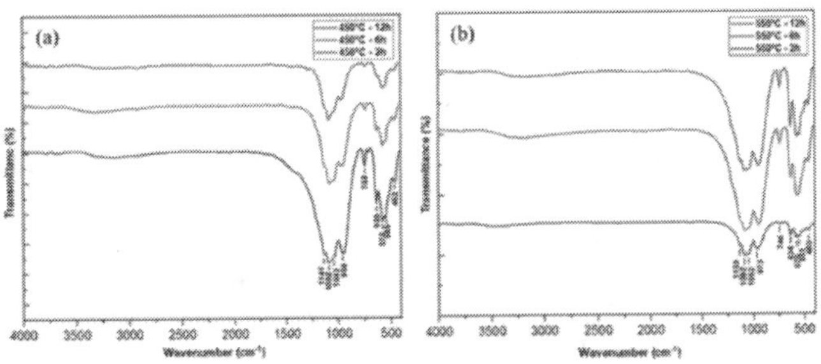

Figure 3. FTIR spectra of $LiMn_{0.5}Fe_{0.5}PO_4$ calcined at (a) 450 °C and (b) 550 °C for 2, 6, and 12 h.

The specific surface area of these $LiMn_{0.5}Fe_{0.5}PO_4$ materials were measured with Brunauer-Emmett-Teller (BET) model and is showed in Table 1. The materials synthesized at 550°C possessed a higher specific surface area. Increasing the reaction time can also increase the surface area which was correlated well with the primary particle size analyses. In other words, the smaller the size of particles, the greater is the specific surface area.

Table 1. Crystallite size, particle size and specific surface area of $LiMn_{0.5}Fe_{0.5}PO_4$ at different conditions.

Temperature (°C)	Time (h)	Crystallite size (nm)	Particle size (nm)	Specific surface area (m^2/g)
	2	24.3	29	5.38
450	6	28.7	24	5.54
	12	29.6	21	6.24
	2	23.4	52	7.31
550	6	26.8	37	12.06
	12	28.3	33	12.36

4. Conclusions

Mn-doped $LiFePO_4$ cathode materials were successfully synthesized by a solid-state reaction method at different temperatures and reaction times. XRD patterns showed that materials for all conditions had identified with the reference data of $LiMn_{0.5}Fe_{0.5}PO_4$, but impurity peaks were found at low synthesis temperatures. The particle became aggregate at high temperature and the smaller particle size was achieved when increasing the reaction time. $LiMn_{0.5}Fe_{0.5}PO_4$ synthesized at 550°C possessed a higher specific surface area than 450°C. The results can provide the optimal conditions for Mn-doped $LiFePO_4$ synthesis, which positively influences the electrochemical performance of the cathode material for Li-ion batteries.

Acknowledgements
This work was financially supported by International Collaborative Education Program for Materials Technology, Education, and Research (ICE-Matter). In addition, the authors gratefully acknowledge the facilities of the Department of Materials Engineering, Faculty of Engineering, Kasetsart University.

References

[1] Wang Y and Yu F 2020 *J. Alloys Compd.* **850** 156773.
[2] Novikova S, Yaroslavtsev S, Rusakov V, Chekannikov A, Kulova T and Skundin A 2015 *J.Power Sources.* **300** 444.
[3] Kang F.-Y, Ma J and Li B.-H 2011 *New Carbon Materials.* **26** 161.
[4] Wang L, Li Y, Wu J, Liang F, Zhang K and Xu R 2020 *J. Alloys Compd.* **839** 155653.
[5] Sun L, Deng Q, Fang B, Li Y, Deng L and Yang B 2016 *CrystEngComm.* **18** 7537.
[6] Lin Y, Zeng B, Lin Y, Li X, Zhao G and Zhou T 2012 *Rare Metals.* **31** 145.
[7] Yaroslavtsev S, Novikova S, Rusakov V, Vostrov N, Kulova T and Skundin A 2018 *Solid State Ionics.* **317** 149.
[8] Hong J, Wang F, Wang X and Graetz J 2011 *J. Power Sources.* **196** 3659.
[9] Saravanan K, Ramar V, Balaya P and Vittal J.-J 2011 *J. Mater. Chem.* **21** 14925.
[10] Yang C.-C, Hung Y.-W and Lue S. J 2016 *J. Power Sources.* **325** 565.
[11] Lu C.-H, Subburaj T, Chiou H.-T, Som S, Ou C and Parthasarathi S. K 2020 *Ionics.* **26** 1051.
[12] Xiang W, Zhong Y.-J, Ji J.-Y, Tang Y, Shen H and Guo X.-D 2015 *Phys. Chem. Chem. Phys.* **17** 18629.
[13] Jo M, Yoo H, Jung Y.-S and Cho J 2012 *J. Power Sources.* **216** 162.
[14] Ruan T, Wang B, Wang F, Song R, Jin F and Zhou Y 2019 *Nanoscale.* **11** 3933.
[15] Burba C. M and Frech R 2007 *J. Power Sources.* **172** 870.

IOP Conf. Series: Materials Science and Engineering 1234 (2022) 012030 doi:10.1088/1757-899X/1234/1/012030

Current generation from graphene vibrations

J Charoenpakdee[1], T Mongkolsawat[1], O Suntijitrungruang[1], and S Boonchui[1,2,*]

[1]Department of Physics, Faculty of Science, Kasetsart University;
[2]Center of Rubber and Polymer Materials in Agriculture and Industry (RPM), Faculty of Science Kasetsart University, Thailand

*Corresponding author: fscistb@ku.ac.th

Abstract. The application of electrical conductivity, extremely flexible graphene membranes, to capture energy from ambient vibrations has promising potential. We study the physics of self-energy from the vibrations and how current is created in graphene for theoretical research. We demonstrate the effects of lattice distortions on the electrical current, such as isotropic tensile strain ϵ_{is}, shear strain ϵ_{ss}, uniaxial armchair strain ϵ_{as}, and zigzag strain ϵ_{zs}. We show how this movement might cause an electrical current to flow in a neighbouring circuit.

1. Introduction

In the early twenty-one century, graphene apparently is a material of this century. Engineers and scientists have researched graphene properties in various aspects like mechanics [1], electronics [2], and optics [3]. As such, one of the interesting researches is the properties of strained graphene. For example, the works [4-7] illustrate Casimir interactions, anisotropic AC conductivity and band-piezoelectric effect in strained graphene respectively. Regarding magnetic properties, the research [7-11] discovers that strained graphene causes the Schwinger pair creation. This research investigates the lattice distortions due to each type of strain such as isotropic tensile strain, shear strain, uniaxial armchair strain, and zigzag strain. Such distortions are an essential factor, hence resulting in certain electromagnetic properties in graphene.

The pseudo magnetic field is one of the effects from the lattice distortions owing to strained graphene. From the theoretical study [7-10], the pseudo magnetic field corresponds to the strains applying on graphene. On this ground, the time variant strains induce the electrical current inside graphene according to Maxwell's equations [12]. Our work will investigate particular electromagnetic characteristics from the lattice distortions in specific strained graphene such as self-energy generation, photon emission distributions and conductivity. The investigation in our study can be applied to improve the electronic properties further as we suggest in the discussion section.

In this paper we present theoretical results of self-energy generation from graphene vibrations. The manuscript is organized as follows. In section 2, we present our theoretical and numerical method for showing self-energy generation as induced electric field and AC current. In section 3, we show the descriptions and the physical conditions for the harmonic emission. Finally, we present our conclusions in section 4.

2. Experimental

Theoretical and numerical method

The implications of deforming the lattice are taken into consideration in our formulation. The hopping will be altered as a result of this. Graphene in our investigation is set up on substrate for aggregation protection. Thus, there is no aggregation in our consideration When the hopping energies between distinct sites are changed, additional terms in the initial Hamiltonian in the tight-binding formulation appear:

$$\hat{H} = -\sum_{\langle ij\rangle} t(|\vec{\delta}_{ij}|)(\hat{a}_i^\dagger \hat{b}_j + \hat{b}_i^\dagger \hat{a}_j) \quad (1)$$

where $\vec{\delta}_{ij}$ is the i th nearest-neighbor vector of the j th atom, and $t(|\vec{\delta}_{ij}|)$ is the strained hopping energy due to strain. The length and direction of the three nearest-neighbor vector $\vec{\delta}_{ij}$ transform under strain according to $\vec{\delta}'_{ij} \approx (1+\vec{\varepsilon})\cdot\vec{\delta}_{i0}$ where $\vec{\delta}_{i0}$ represents a nearest-neighbor vector in the non-deformed graphene plane, and $\vec{\varepsilon}$ is the strain tensor [7,8]. In Cartesian coordinate system, we assume the x-axis is along graphene armchair direction and the y-axis is along graphene's zigzag (Figure 1). In reciprocal space, the variation of hopping energy with inter carbon distance is $t(|\vec{\delta}_{ij}|) = t_0 \exp[-\beta(|\vec{\delta}_{ij}|/a-1)]$ with $a = 1.42\ A^0$ being the unstrained nearest-neighbor separation, $t_0 = 2.7\ eV$, and $\beta = 3$ [7,8]. The hopping perturbation δti can be obtained by expanding $t(|\vec{\delta}_{ij}|)$ and keeping the first order:

$$t(|a+\Delta\vec{\delta}_{ij}|) \approx t_0(1-(\beta/a)|\vec{\delta}'_{ij} \approx (1+\vec{\varepsilon})\cdot\vec{\delta}_{i0}| - |\vec{\delta}_{i0}|] \quad (2)$$

with $\Delta\vec{\delta}_{ij}$ being the variation of bond length.

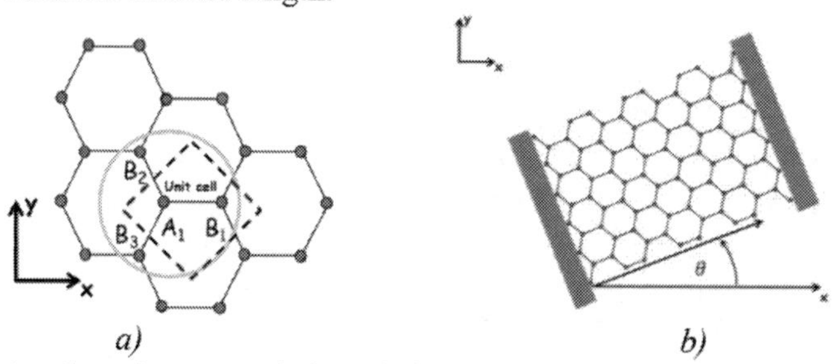

Figure 1. a) Schematic of graphene armchair and zigzag directions. b) Illustration of strained graphene with angle θ made by the direction of extensional strain and the x-axis.

Following ref. [7-10], we obtain the Hamiltonian in the reciprocal space near the K points as

$$H = \sum_{\mu=1,2} \hbar\vec{v}_\mu \cdot \vec{k}\sigma^\mu + (e/c)\vec{A}_{pse}\cdot\vec{\sigma} \quad (3)$$

It is the form of the Weyl–Dirac equation for the deformed graphene where σ^μ are the Pauli matrices. The 2D-velocities $\vec{v}_\mu = (v_\mu^x, v_\mu^y)$ is defined by its components as

$$v_1^x = -v_F(\frac{1}{2}\beta\varepsilon_{xy} - \frac{2\pi}{3\sqrt{3}}\varepsilon_{yx}),\quad v_1^y = v_F(1-\frac{1}{4}\beta(\varepsilon_{xx}+3\varepsilon_{yy}) - \frac{2\pi}{3\sqrt{3}}\varepsilon_{yy}) \quad (4a)$$

and

$$v_2^x = v_F(1-\frac{1}{4}\beta(3\varepsilon_{xx}+\varepsilon_{yy}) + \frac{2\pi}{3\sqrt{3}}\varepsilon_{yy}),\quad v_2^y = -v_F(\frac{1}{4}\beta(\varepsilon_{xy} - \frac{2\pi}{3\sqrt{3}}\varepsilon_{yx})). \quad (4b)$$

As we have seen, the Weyl–Dirac Hamiltonian is described by the velocities $\vec{v}_1$ and $\vec{v}_2$ which are defined in terms of four elements of the strain tensors, ε_{xx}, ε_{xy}, ε_{yx} and ε_{yy}. These are defined as

$$\vec{\varepsilon} = \begin{pmatrix} \varepsilon_{xx} & \varepsilon_{xy} \\ \varepsilon_{yx} & \varepsilon_{yy} \end{pmatrix} = \varepsilon \begin{pmatrix} \cos^2\theta - \nu\sin^2\theta & (1+\nu)\sin\theta\cos\theta \\ (1+\nu)\sin\theta\cos\theta & \sin^2\theta - \nu\cos^2\theta \end{pmatrix}, \qquad (5)$$

where ε is the strain modulus, ν is the Poisson's ratio and θ is an angle made by the direction of the principal strain and the x-axis illustrated in Figure 1 b).

The pseudo-vector potential $\vec{A}_{pse}$ is given by

$$\vec{A}_{pse} = v_F\left(\frac{c\pi\hbar}{ea}\right)\left(\left(\frac{4}{3\sqrt{3}}\varepsilon_{yy} + \frac{\beta}{2\pi}(\varepsilon_{xx}+\varepsilon_{yy})\right), \left(\frac{4}{3\sqrt{3}}\varepsilon_{yy} + \frac{\beta}{2\pi}\varepsilon_{yx}\right)\right). \qquad (6)$$

In case of uniform strain, the pseudo-vector potential appeared in the Weyl-Dirac equation is uniform and then there is no pseudo-magnetic field present ($\vec{B}_{pse} = \vec{\nabla}\times\vec{A}_{pse} = 0$). For graphene vibrations, ε the strain modulus is given $\varepsilon = \varepsilon_0 \sin(\omega t)$ and ε_0 is an amplitude of strain modulus and ω is a frequency of vibration. Then the pseudo-electric field can be $\vec{E}_{pse} = -\partial\vec{A}_{pse}/\partial t$.

Let us now obtain the AC conductivity $\sigma_{\alpha\beta}$ of graphene under uniform strain by combining (3) and the Kubo formula, assuming a linear response to an external electric field with frequency w. Following the representation used in [12], the current density operators j_α can be written as

$$j_x = -ie[\hat{H},\hat{x}] = e\partial\hat{H}/\partial q_x \approx (1-\tilde{\beta}\varepsilon_{xx})j_x^* - \tilde{\beta}\varepsilon_{xy}j_y^*, \qquad (7a)$$

and

$$j_y = -ie[\hat{H},\hat{y}] = e\partial\hat{H}/\partial q_y \approx (1-\tilde{\beta}\varepsilon_{yy})j_y^* - \tilde{\beta}\varepsilon_{xy}j_x^*, \quad \tilde{\beta} = (\beta-1) \qquad (7b)$$

where $j_\alpha^* = \sigma_{\alpha,\lambda}^* E_\lambda$ are the current operator components ($\alpha, \lambda = x(y)$) for unstrained graphene.

3. Results and discussion

Self-energy generation as induced electric field and AC current.

Considering Figure 2.a), the simulations show that the pseudo-vector potential in x direction is the same phase, despite the different angles. From Eq. (6), we find that the pseudo-vector potential in x direction is derived from the strain tensor in diagonal elements that are normal strain. Regarding Figure 2.b, the tendency of the phase (for pseudo-vector potential in y directions) is almost similar to the tendency for x directions. Nevertheless, in the case of zero angle, the phase of this angle is apparently opposite to the phase of the other angels. Through theoretical analysis, the zero angle causes the pseudo-vector potential in y directions depending on strain in element *yy* only.

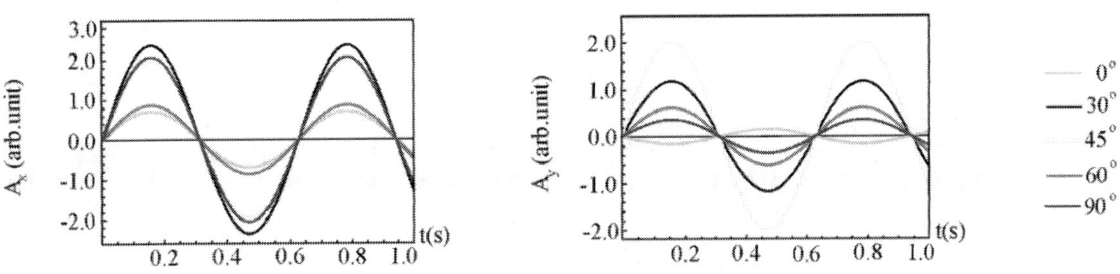

a) b)

Figure 2. Schematic of the pseudo-vector potential in Eq.(6) for a) x directions and b) y directions.

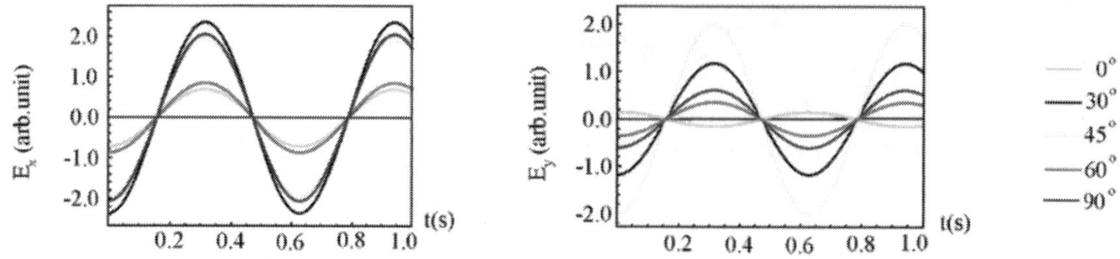

a) b)

Figure 3. Schematic of the pseudo electric field in Eq.(6) for a) x directions and b) y directions.

According to the reference [7,8], the zero angle makes the strain to be uniaxial armchair type. Next, considering the pseudo electric field shown in Figure 3. All of these graphs have the phase that shifts from Figure 2 equal to -90° owing to Maxwell's equation.

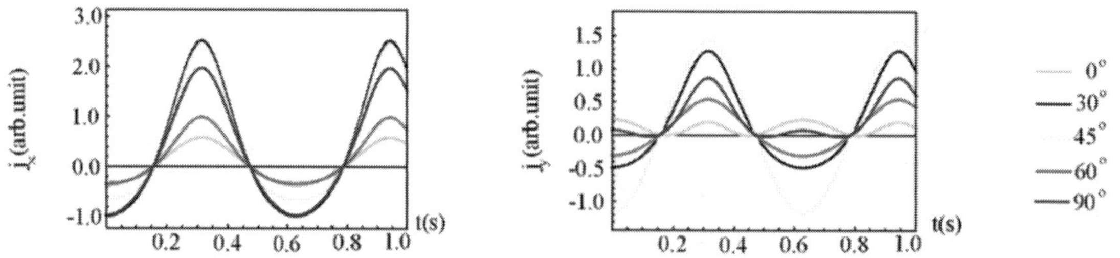

a) b)

Figure 4. Schematic of the current in Eq.(7a) and (7b) for a) x directions and b) y directions.

Other than the phase, all tendencies of each graph are still the same. Then, to observe the current characteristic in strained graphene, we have to consider Figure 4. Looking at this figure, we discover that the phase of the current is the same as the pseudo electric field shown in Figure 3. Considering theoretical detail, Eq.7 (both in a and b) implies that the current's phase is consistent with the pseudo electric fields.

4.Conclusions

Our research demonstrates that the strain induces the pseudo-vector potential inside graphene. With the time dependent strain, the pseudo electric field and the current take place in graphene according to the Maxwell equation. We also discover that the different strains make the different current characteristics. This means that the conductivity of strain graphene can be adjusted by the strains applying on graphene. Therefore, based on our work, it provides a way to determine graphene's conductivity by applying specific strains.

Acknowledgements

We acknowledge helpful discussions with Philip (Haiyong) for discussions. This work was supports by The Royal Golden Jubilee (RGJ) Ph.D. Programme and Faculty of Science Kasetsart University for partial support.

References

[1] Lee, Changgu Science. 321, 385 (2008) doi:10.1126/science.1157996.
[2] Cooper, Daniel R.; et al. , ISRN Condensed Matter Physics 1–56 (2012).
[3] Nair, R. R.; Blake, P.; Grigorenko, A. N.; Novoselov, K. S.; Booth, T. J.; Stauber, T.; Peres, N. M. R.; Geim, A. K. Science. 320 (5881): 1308 (2008).

[4] Anh D. Phan and The-Long Phan, Phys. Status Solidi RRL 8, No. 12, 1003–1006 (2014).
[5] M. Oliva-Leyva and Gerardo G. Naumis, J. Phys.: Condens. Matter 26 125302 (2014).
[6] Wang, X., Tian, H., Xie, W. et al., NPG Asia Mater 7, e154 (2015).
[7] P. Fanbanrai, A. Hutem, S. Boonchui, Physica B: Condensed Matter. Vol. 472 (2015) Pages 84-90.
[8] S. Boonchui and P. Nualpijit, Opt. Express 27, 28350-28363 (2019).
[9] Thitapura, T., Liewrian, W., Jutarosaga, T. et al., Plasmonics 12, 1439-14475 (2017).
[10] Thiti Thitapura, Watchara Liewrian, Tula Jutarosaga, and S. Boonchui, Opt. Express 25, 25588-25601 (2017).
[11] Alexander L. Kitt, Vitor M. Pereira, Anna K. Swan, and Bennett B. Goldberg, Phys. Rev. B 87, 159909 (2013).
[12] J. D. Jackson (1975-10-17). Classical Electrodynamics (3rd ed.). ISBN 978-0-471-43132-9.e
[13] J. Charoenpakdee, , O. Suntijitrungruang, & S. Boonchui, Sci Rep 10, 18949 (2020).

Valorization of rubberwood waste into porous carbon

R Songthan[1], V Somsongkul[1,2]*, and C Kongmark[3]**

[1] Department of Industrial Chemistry, Faculty of Applied Science, King Mongkut's University of Technology North Bangkok, Bangkok 10800, Thailand
[2] Integrated Chemistry Research Center for Sustainable Technology (IRCT), King Mongkut's University of Technology North Bangkok, Bangkok 10800, Thailand
[3] Specialized Center of Rubber and Polymer Materials in Agriculture and Industry (RPM), Department of Materials Science, Faculty of Science, Kasetsart University, Bangkok 10900, Thailand

*Corresponding author email address: voranuch.s@sci.kmutnb.ac.th
**Corresponding author email address: chanapa.k@ku.th

Abstract. Rubberwood is one of the main Thailand goods exports. In each year, about 45,000 hectares of rubber tree are cut down, only about 30% of them is appropriated for furniture production while most of the rest is used as woodfuel. This research work focuses on the valorization of rubberwood waste into porous carbon and the optimization of carbonization process for the production of hierarchical porous carbon. Rubberwood wastes obtained from a rubberwood processing industry (in Thailand's southern region) were treated with H_3PO_4 and subjected to two different carbonization methods. In the first method, the rubberwood was carbonized at 500°C for 1 h (PC-1step). The second method consisted of two steps of heat treatment, the rubberwood was pre-carbonized at 200°C for 15 min and subsequently carbonized at 500°C for 45 min (PC-2step). The physical properties of the porous carbon were investigated using thermogravimetric analysis, Fourier transform infrared spectroscopy, scanning electron microscopy, and N_2 adsorption-desorption isotherm techniques. The porous carbons obtained from both carbonization methods have hierarchical porous structures comprising the naturally existing tubular macropores (diameter 4-20 μm) and the micropores (diameter 0.3-1 nm) generated during carbonization process. The PC-2step provides a specific surface area of 336.2 $m^2\,g^{-1}$ and a pore volume of 0.093 $cm^3\,g^{-1}$, which are almost twice those of the PC-1step. It is possible that the pre-carbonization at 200°C would allow a slow and gradual evaporation of moisture within the wood, thus preserving pore walls in the materials. The obtained hierarchical porous carbons can be used in many applications such as energy storage, waste water treatment, air purification and metal recovery.

1. Introduction

The world's top producers and exporters of natural rubber are Southeast Asia (SEA) countries, particularly Thailand, Malaysia, and Indonesia. Rubberwood plantations in Thailand covers an area of approximately 20 million square meters by 2020 [1]. The rubber trees generally can be harvested for about 20-30 years before their latex production starts to slow down. And after that, rubber trees are cut down and transported to the wood processing industry, where only about 30% of them are used for furniture production, where a large amount of rubberwood waste is generated, and most of the rest are used as woodfuel. This research work focuses on the valorization of rubberwood wastes into porous carbon.

The hierarchical porous carbons can be used in many applications, such as energy storage, waste water treatment, air purification, and metal recovery [2]. The carbonization process plays key role for the production of porous carbon from biomass. It also makes use of chemical activation to increase the porosity of porous carbon materials. The common chemical agents are H_3PO_4, $ZnCl_2$, KOH, and NaOH. Among them, phosphoric acid has proven to be the most suitable for wood biomass activation since it is highly effective, cost-effective, and environmentally friendly [3]. Taer *et al.* [4] reported in 2011 that porous carbon derived from rubberwood sawdust had a high surface area (263-684 $m^2\ g^{-1}$). Later in 2017, Thubsuang *et al.* [5] obtained porous carbon from rubberwood waste treated with H_3PO_4, which exhibited a large surface area (605-693 $m^2\ g^{-1}$).

The purpose of this work is to study the carbonization process to transform the rubberwood waste into porous carbon with high porosity, high surface area, and high pore volume which can be used in many applications such as energy storage, wastewater treatment, air purification and metal recovery.

2. Experimental

Rubberwood waste was obtained from a rubberwood processing industry (in Thailand's southern region) and was utilized as a raw material in the production of porous carbon samples. Phosphoric acid (H_3PO_4, 85%) was purchased from Qrëc.

Rubberwood waste was cut into cubes of 1 cm^3. Subsequently, they were dried at 100°C for 24 h. The dried rubberwood samples of about 30 g were activated by 60% w/w H_3PO_4, with an impregnation ratio of 2:1 (i.e., mass ratio of H_3PO_4 to wood) for 24 h [5]. Afterwards, the samples were dried at 100°C for 24 h. There are two different carbonization methods. In the first method, the carbonization was achieved at 500°C for 1 h, with a heating rate of 5°C min^{-1} under a nitrogen flow of 100 mL min^{-1}. The second method consisted of two steps of heat treatment, the rubberwood was pre-carbonized at 200°C for 15 min and subsequently carbonized at 500°C for 45 min. Finally, the porous carbon was washed with distilled water and dried at 100°C for 12 h.

The thermal stability was studied by thermogravimetric analysis-TGA (TGA/DSC1, Mettler Toledo) in the temperature range 30-900°C with a heating rate of 10°C min^{-1} under a nitrogen flow of 100 mL min^{-1}. The functional groups in samples were analyzed by Fourier transform infrared spectroscopy-FTIR (Vertex 70, Bruker) in the range of 200-4000 cm^{-1}. The morphology of samples was studied by Scanning Electron Microscopy-SEM (QUANTA 450 FEI; Oxford instrument). Nitrogen adsorption-desorption measurement was used to determine the pore structure and textural properties of materials (3Flex surface characterization, Micromeritics). The specific surface area (S_{BET}) was calculated using Brunauer-Emmett-Teller (BET) method. The micropore volume (V_{micro}) was calculated using t-plot method. The mesopore volume (V_{meso}) was calculated using BJH method. The pore size distribution and average pore size were analyzed using the Hovath-Kawazoe (HK) models.

3. Results and discussion

Figure 1a shows the TGA and DTG curves of the rubberwood. Rubberwood shows three thermal decomposition zones: 30-200°C, 200-400°C and 400-900°C. The first weight loss of around 4.71% is related to the dehydration of moisture in rubberwood. The second weight loss of 62.92% is related to the transformation of hemicellulose and cellulose into non-condensable (primarily CO, CO_2, H_2 and CH_4) gas and condensable organic vapours. The last weight loss of 16.95% is possibly due to the decomposition of a structure with higher stability, such as lignin [6]. Therefore, the carbonization process in the present study was carried out at 500°C.

Figure 1b shows the FT-IR spectra of rubberwood and porous carbon. The FT-IR spectra of rubberwood shows peaks at a wavenumber of 3338 and 1623 cm^{-1} corresponding to the O-H stretching and bending of moisture in rubberwood [7-8]. The peaks at 2916 and 1369 cm^{-1} are attributed to the C-H stretching and bending of lignocellulose [9-10]. The peaks at 1421 and 1456 cm^{-1} correspond to the C-H deformation in lignin [7,10]. The bands at 1593 and 1504 cm^{-1} belong to the C=C stretching vibrations of lignin. The bands at 1731 and 1236cm^{-1} correspond to C=O stretching and C-O bending of hemicellulose and lignin [10-11]. A strong peak at 1022 cm^{-1} is attributed to the C–O–C pyranose ring of lignocellulose [7]. Most of these peaks were not observed in porous carbons due to the decomposition of lignocellulose into gas and tar. However, small characteristic IR bands (1558 and 1124 cm^{-1}) of lignin are still present in porous carbons, which are possibly due to the wide decomposition temperature range of lignin (150-900°C) [9,11].

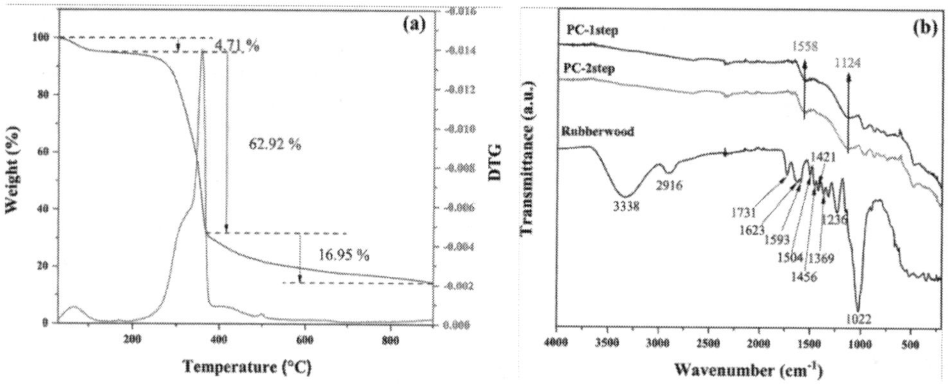

Figure 1. (a) TGA curves of rubberwood and (b) FT-IR spectra of rubberwood and porous carbons.

Figure 2 shows SEM images of rubberwood and porous carbons. Rubberwood exhibits tubular structure resemble to the xylem. Porous carbons show pores similar to a honeycomb structure. The porous carbons obtained from both carbonization methods have hierarchical porous structures comprising the naturally existing tubular macropores (diameter 4-20 µm) and the micropores generated during carbonization process due to the decomposition of moisture and volatile matter into tars (condensed volatiles) and gas.

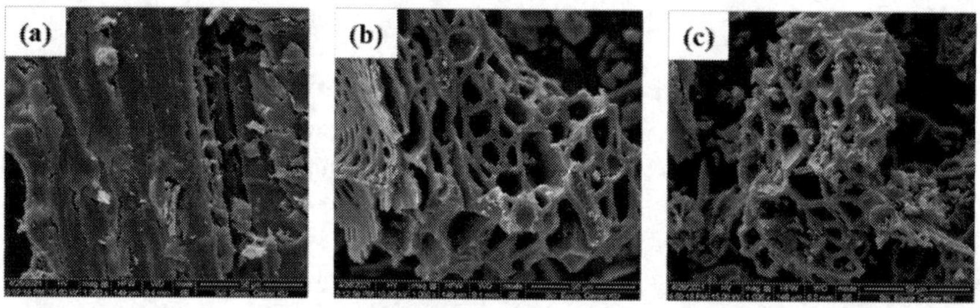

Figure 2. SEM images of (a) rubberwood, (b) PC-1step, and (c) PC-2step.

Figure 3a presents the nitrogen adsorption isotherm of PC-1step and PC-2step. According to the classification by IUPAC, the isotherms for all of the samples are found to be type-I characteristics [12]. It is obvious that the slope of the isotherm at a relative pressure range of 0.01-0.03 of PC-2step is higher than that of PC-1step, which would indicate that PC-2step has greater quantities of micropores than PC-1step. Figure 3b shows the pore size distribution of PC-1step and PC-2step, which are in the range of 0.3 nm to 0.7 nm. The average pore sizes of PC-1step and PC-2step are approximately the same at 0.44 nm. Table 1 shows the textual properties of porous carbons. The PC-2step provides a specific surface area of 336.2 m^2 g^{-1} and pore volume of 0.093 cm^3 g^{-1}, which are almost twice those of the PC-1step. It is possible that the pre-carbonization at 200°C would allow a slow and gradual evaporation of moisture within the wood, thus preserving pore walls in the materials [13]. Furthermore, at temperatures above 217°C, pyrophosphoric acid ($H_4P_2O_7$) can be formed by the condensation of two or more H_3PO_4 molecules with the elimination of H_2O. The pyrophosphoric acid has a strong corrosivity, which may enhance surface area and pore volume in the PC-2 step [14]. Although the present study has successfully demonstrated the importance of the pre-carbonization process which enhances the surface area and porosity of carbon, the obtained porous carbons have a relatively low surface area compared to the commercial carbon, e.g., YP50F with a surface area of 1410 m^2 g^{-1} [15]. In fact, there are several factors affecting the properties of porous carbon, including carbonization time, carbonization temperature, activating agent, and mass ratio of activating agent to wood, which will be investigated in the future.

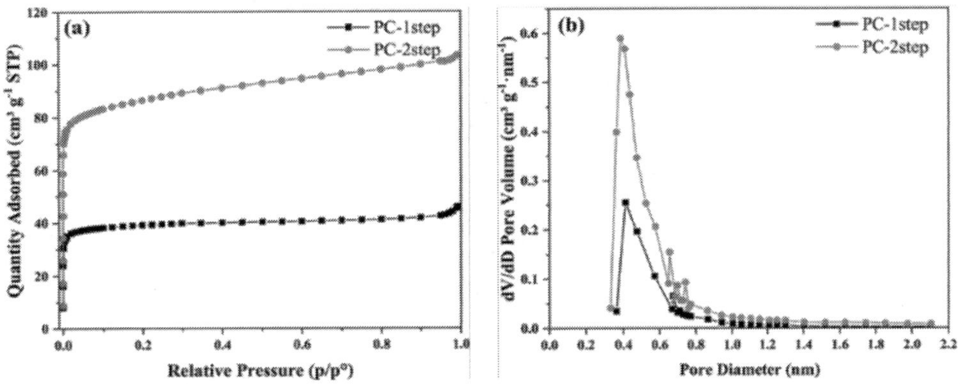

Figure 3 (a) Nitrogen adsorption-desorption isotherm (b) micropore size distribution of porous carbons.

Table 1. The physical properties of porous carbons obtained from rubberwood treated with H_3PO_4.

Sample	S_{BET} (m^2 g^{-1})	S_{micro} (m^2 g^{-1})	V_{total} (cm^3 g^{-1})	V_{meso} (cm^3 g^{-1})	V_{micro} (cm^3 g^{-1})	D (nm)
PC-1step	156.6	121.3	0.046	0.015	0.071	0.44
PC-2step	336.2	239.2	0.093	0.046	0.159	0.44

4. Conclusions

Both carbonization procedures produce porous carbons with hierarchical porous structures composed of tubular macropores (diameter 4-20 μm) and micropores (diameter 0.3-1 nm). Porous carbon prepared by pre-carbonization at 200°C and subsequently carbonization at 500°C provided a specific surface area of 336.2 m^2 g^{-1} and a pore volume of 0.093 cm^3 g^{-1}, which are almost twice that of the porous carbon carbonized only at 500°C. It is possible that the high porosity of porous carbon resulted from the gradual evaporation of moisture within the wood during pre-carbonization process at 200°C and the chemical corrosion caused by phosphorous species.

Acknowledgement

This project is funded by National Research Council of Thailand (NRCT), the Faculty of Science, Kasetsart University and the Faculty of Applied Science, King Mongkut's University of Technology North Bangkok (Contract no. 6141108).

References

[1] https://www.raot.co.th/ewt_dl_link.php?nid=6750: October 2021.
[2] Frackowiak E and Béguin F 2001 *Carbon* **39** 937.
[3] Danish M and Ahmad T 2018 *Renew. Sust. Energ. Rev.* **87** 1.
[4] Taer E, Deraman M, Talib IA, Umar AA, Oyama M and Yunus RM 2010 *Curr. Appl. Phys.* **10** 1071.
[5] Thubsuang U, Laebang S, Manmuanpom N, Wongkasemjit S and Chaisuwan T 2017 *J. Mater. Sci.* **52** 6837.

[6] Shaaban A, Se SM, Mitan NM and Dimin MF 2013 *Procedia Eng.* **68** 365.
[7] Nazarpour F, Abdullah DK, Abdullah N and Zamiri R 2013 *Materials* **6** 2059.
[8] Sun XF, Sun RC, Tomkinson J and Baird MS 2004 *Polym. Degrad. Stab.* **83** 47.
[9] Solikhin A, Hadi YS, Massijaya MY and Nikmatin S 2019 *Waste and Biomass Valorization* **10** 953.
[10] Le Troedec M, Sedan D, Peyratout C, Bonnet JP, Smith A, Guinebretiere R, Gloaguen V and Krausz P 2008 Compos. *Part A Appl. Sci.* **39** 514.
[11] Pandey KK and Pitman AJ 2003 Int. *Biodeterior. Biodegradation* **52** 151.
[12] Donohue MD and Aranovich GL 1998 *Adv. Colloid Interface Sci.* **76** 137.
[13] Srinivasakannan C and Bakar MZA 2004 *Biomass Bioenergy* **27** 89.
[14] Liu H, Zhang J, Bao N, Cheng C, Ren L and Zhang C 2012 *J. Hazard. Mater.* **235** 367.
[15] Yang D, Jing H, Wang Z, Li J, Hu M, Lv R, Zhang R and Chen D 2018 *J. Colloid Interface Sci.* **528** 208.

IOP Conf. Series: Materials Science and Engineering 1234 (2022) 012032 doi:10.1088/1757-899X/1234/1/012032

Mechanical properties of laser-welded and adhesive-bonded ultra-high-strength steel lap joints

M Hietala[1*], M Keskitalo[1] and A Järvenpää[1]
[1] University of Oulu, Kerttu Saalasti Institute, FI-85500 Pajatie 5 Nivala, Finland

*Corresponding author: mikko.hietala@oulu.fi

Abstract. The need to manufacture lightweight structures from ultra-high-strength steels is forcing industry to look for new joining technologies. Hybrid joints are one of the new ways to join materials. This paper investigates joining ultra-high strength steel plates using laser welding and adhesive bonding. Steel used in this paper was abrasion resistant steel (AR600) with tensile strength (Rm) $\geq$ 2 GPa. Three different epoxy adhesives were used in the experiments. The weld morphologies were investigated using optical microscopy. The mechanical properties of the lap joints were evaluated by hardness measurements and tensile shear tests. Examining the profiles of the joints, it became clear that the adhesive causes an air gap between the plates, which affects the morphology of the joint. The air gap caused by the adhesive increased the width of the weld at the interface of the plates. The width of the weld at the interface was approximately 25% larger in the adhesive-bonded joint. Based on shear strength tests, the use of adhesive in addition to the laser weld significantly improved the shear strength of the lap joint. The shear strength of the best laser-welded and adhesive-bonded joint was approximately 84% higher than that of laser-welded joint.

1. Introduction

Due to the reduction of traffic emissions, new ways of joining steels are needed in new vehicle structures. New assembly technologies have received a lot of interest lately and have grown in popularity at the expense of more traditional welding methods. The strength of the joints must be as good as possible so that the properties of the ultra-high-strength steels can be utilized in different structures.

Laser welding and adhesive bonding are widely used in joints of various vehicle structures [1,2]. Laser welding and adhesive bonding are often considered to be competing joining methods. However, there are structures where laser welding and adhesive bonding can be utilized together. Laser welding is excellent method for joining sheet metal products [3]. The advantage of laser welding is a very low heat input which is why the heat-affected-zone (HAZ) of the laser weld is very narrow [4-6]. Due to the low heat input, laser welding is also suitable for use with adhesive joints, as the mechanical properties of the adhesive joint do not change under the influence of heat.
Wang et al. investigated influence of adhesives in laser weld bonded joints [7]. Their observation was that adhesive improved the mechanical properties of the laser welded joint. Wang *el al.* studied laser-arc-adhesive hybrid welding [8]. They found that Mg/Al dissimilar metals were successfully joined by laser-arc-adhesive welding technology.

IOP Conf. Series: Materials Science and Engineering 1234 (2022) 012032 doi:10.1088/1757-899X/1234/1/012032

At the present time, there are very few papers with results reported on the mechanical properties of the laser-welded and adhesive-bonded joints. The objective of the paper was to study mechanical properties of laser-welded and adhesive-bonded lap joints and to determine which of the studied adhesives achieves the highest shear strength combined with a laser weld. The morphologies of the joints were investigated by optical microscopy. The microhardness and shear strengths of the lap joints were investigated.

2. Experimental methods

2.1. Test material and adhesives used in experiments

The test material used in experiments was ultra-high-strength martensitic abrasion resistant steel (AR600). The thickness of the AR600 was 3 mm. The measured mechanical properties of the AR600 are shown in table 1. Three different structural bonding epoxy adhesives were used in the experiments. All adhesives were two-component adhesives. Adhesives were Loctite 9466 (Adhesive 1), SikaPower 4720 (Adhesive 2) and SikaPower 1277 (Adhesive 3).

Table 1. The measured mechanical properties of AR600.

Material	Yield strength (MPa)	Tensile strength (MPa)	Elongation (%)	Hardness (HV0.2)
AR600	1774	2164	6.4	646

2.2. Experimental setup

The test specimens were manufactured by using laser cutting and the dimensions of the single specimen were 200 mm × 60 mm × 3 mm. Dimensions of the laser welded specimen are shown in Figure 1. An overlap size of the sheets was 100 × 60 mm. There was 20 mm gap between lap joint and adhesive so that the properties of the adhesive were not affected by the laser weld. Adhesive-bonded area was 3600 mm². The plates were adhesive-bonded first, and laser welded after.

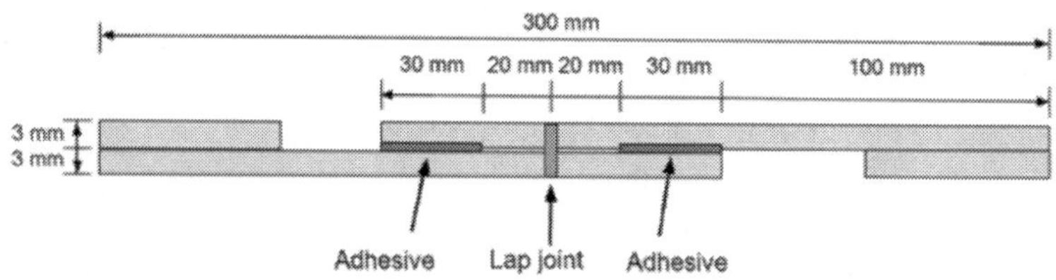

Figure 1. Dimensions of the test specimen.

Trumpf HLD-4002 continuous mode Yb:YAG disc-laser with a laser wavelength of 1030 nm was used in laser welding experiments. Fixed power 4 kW and welding speed 1.5 m/min was used. The energy input of the weld was 160 J/mm. Focus was adjusted 1 mm below of the plate surface. Laser beam diameter was 0.3 mm on the surface of the plate. Welding optics focal length was 300 mm.

2.3. Mechanical characterization

The tensile shear strength of the laser-welded lap joints was evaluated according to the standard ASTM E8/8M by carrying out tensile tests at a constant loading rate of 1.0 mm/min using Instron 8802 universal material testing machine. Tensile shear strength tests were made in the direction perpendicular to the lap joint. Five tensile shear strength tests were performed for each lap joint and the average result was reported. The Vickers hardness of the joint was measured in the weld cross-section using 0.2 kg load. The hardness was measured every 0.1 mm distance.

3. Results and discussion

3.1. Weld morphology

The morphologies of the joints are shown in Figure 2. As shown in Figure 2a and b, the morphology of the joint is affected by the air gap caused by the adhesive. The adhesive created an air gap of approximately 0.3 mm between the plates. The width of the weld at the interface of the plates was 1.12 mm when there was no adhesive between the plates. The width of the weld at the interface of the plates increased when there was adhesive between the plates and as can be seen in Figure 2b, the width of the joint was 1.39 mm. The air gap also had an effect on the width of the weld face as well as the penetration of the weld.

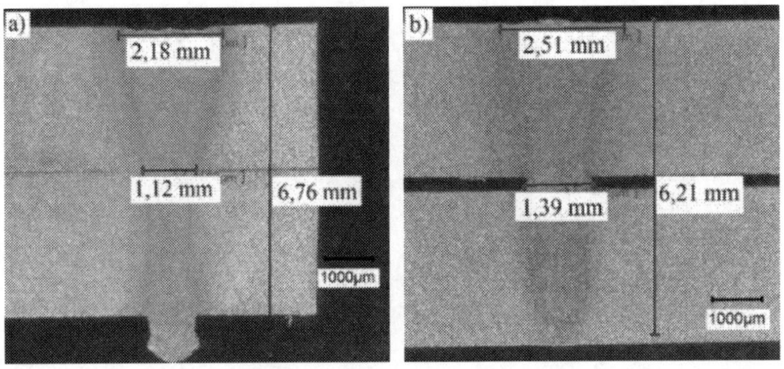

Figure 2. Morphology of laser-welded lap joints with energy input of 160 J/mm (a) 0 mm air gap (b) 0.3 mm air gap.

3.2. Microhardness of the laser weld

The microhardness profile of the lap joint can be seen in Figure 3a. The hardness of the AR600 base material was 646 HV. With energy input used in experiments the hardness of the weld metal was approximately 598 HV. There was softened zone in the HAZ of the joint. Minimum hardness of the HAZ was 371 HV. Width of the HAZ was approximately 9.2 mm. It can be concluded from the width of the HAZ that the thermal effect of the laser weld did not extend to the adhesive-bonded area.

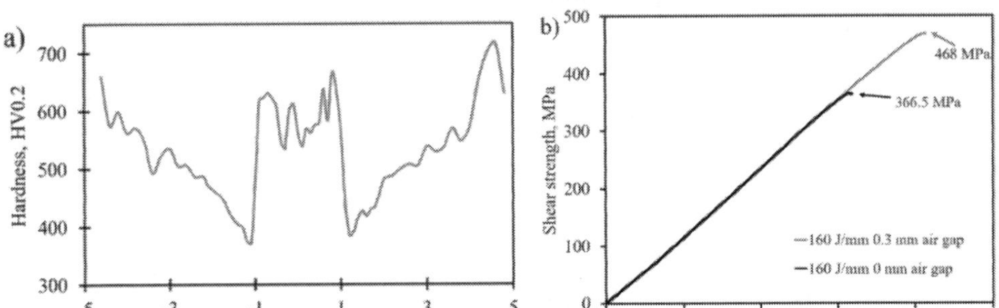

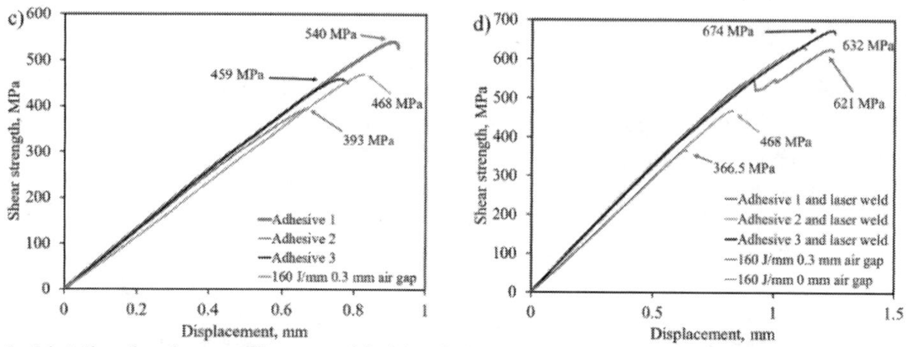

Figure 3. (a) Microhardness of laser-welded lap joint, (b) Shear strengths of laser-welded lap joints, (c) Shear strengths of adhesive-bonded lap joints, (d) Shear strengths of laser-welded and adhesive-bonded lap joints.

3.3. Shear strength of the lap joints

It was investigated how the air gap caused by the adhesive affected the shear strength of the laser-welded lap joint. As seen in Figure 3b. the shear strength of the lap joint increased 28% as the air gap increased from 0 to 0.3 mm. The increase in the strength of the lap joint is due to the increase in the width of the weld at the interface of the plates caused by the air gap.

Shear strengths of the adhesive-bonded joints were tested under tensile shear mode. The peak load and corresponding displacement are shown in Figure 3c. The shear strengths of the adhesive-bonded joints were 393, 459, and 540 MPa for Adhesive 2, Adhesive 3 and Adhesive 1, respectively.

The shear strengths of the laser-welded and adhesive-bonded (LW-AB) joints are shown in Figure 3d. The shear strength of all LW-AB joints was significantly higher than that of the laser-welded joint. the highest joint strength was achieved with laser-weld and adhesive 3 with a shear strength of 674 MPa. Adhesive 1 achieved the highest shear strength in pure adhesive-bonded joints but was not the best adhesive in LW-AB joints. This was because the other half of the adhesive-bonding fractured before the final fracture of the joint occurred as seen from Figure 3d. This is due to the brittleness of the Adhesive 1. It was observed that the shear strength of the LW-AB joint was significantly higher compared to laser-welded lap joint. The strength of the LW-AB joint was 84% higher with Adhesive 3 than that of laser-welded joint with an air gap of 0 mm.

4. Conclusions

Laser-welded and adhesive-bonded lap joints of ultra-high-strength abrasion resistant steel (AR600) were investigated. Weld morphologies, microhardness and tensile shear strength of the lap joints were evaluated. The following conclusions can be drawn:

- The adhesive created an air gap of approximately 0.3 mm between the plates. The air gap caused by the adhesive increased the width of the weld at the interface of the plates. The width of the weld at the interface was approximately 25% larger in the adhesive-bonded joint. The shear strength of the lap joint increased 28% as the air gap increased from 0 to 0.3 mm.
- The hardness of the weld metal was approximately 598 HV. There was clear softened zone in the HAZ of the joint. Minimum hardness of the HAZ was 371 HV.
- The shear strength of the laser-welded and adhesive-bonded joints was significantly higher compared to laser-welded lap joint. The shear strength of the best laser-welded and adhesive-bonded joint was approximately 84% higher than that of laser-welded joint.

Acknowledgements
The authors would like to acknowledge the financial support received from the European Union (European regional development fund), City of Nivala, Nivala industrial park Ltd, NIHAK Nivala-Haapajarvi region registered association. The industrial companies Wärtsilä Finland Oyj, SSAB Europe Oyj, HT Laser Oy, Randax Oy and Miilux Oy have participated in this research work.

References
[1] Bartczak B, Musca J and Trzepieciński T 2013 Int. J. Adhes. Adhes. **45** 42.
[2] Wu Q, Gong J, Chen G and Xu L 2008 Opt. Laser Technol. **40** 420.
[3] Mosavi A, Soleimani A, Karimi A, Akbari M, Karimipour A and Karimipour A 2020 Infrared. Phys. Technol. **111** 103478.
[4] Meng W, Li Z, Huang J, Wu Y and S. Katayama 2014 J. Mater. Eng. Perform. **23** 538.
[5] Grajcar A, Matter P and Stano S 2017 J. Mater. Eng. Perform. **26** 1920.
[6] Kang M, Jeon I H, Han H N and Kim C 2018 Metals. **8** 365.
[7] Wang H Y and Liu L M 2014 Int. J. Adhes. Adhes. **52** 77.
[8] Wang H, Liu L and Liu F 2013 Mater. Des. **50** 463.

IOP Conf. Series: Materials Science and Engineering 1234 (2022) 012033 doi:10.1088/1757-899X/1234/1/012033

High-harmonic generation (HHG) in silicene

C Summueang[1], T Fudulwatjananon[1], and S Boonchui[1,2,*]
[1]Department of Physics, Faculty of Science, Kasetsart University;
[2]Center of Rubber and Polymer Materials in Agriculture and Industry (RPM), Faculty of Science Kasetsart University, Thailand

*Corresponding author: fscistb@ku.ac.th

Abstract. Silicene, a freshly isolated silicon allotrope with a two-dimensional (2D) honeycomb lattice structure, is expected to have electrical properties comparable to graphene. Considering the certain external electric applying on silicene, we explore high-harmonic generation (HHG) effect and its factors. According to our investigation, the external electric field significantly influences the optical emission peaks of the low-frequency optical emission.

1. Introduction

The novel two-dimensional material (such as graphene) and carbon nanotube have considerably attracted scientists both in theoretical and experimental recently [1-6], due to its exotic electronic structure and effective applications in nanoelectronics as well as electrical and optical properties. Silicene, silicon's equivalent of graphene, is usually created by epitaxial growth with featured a slightly bowed honeycomb structure [7], making one of the most promising option for optoelectronic applications. For a strong laser pulse illuminates two-dimensional mater. Under these circumstances, the sample will emit the generating beam's high harmonics (above the fifth harmonic). High harmonics generation (HHG) is a non-linear process due to the coherent nature of the process, HHG is a prerequisite of attosecond physics.

In this paper we present theoretical results of HHG in monolayer silicene film induced by few-cycle laser pulses at infrared wavelengths. The manuscript is organized as follows. In section 2, we present our theoretical and numerical method for showing HHG spectra. In section 3, we show the descriptions and the physical conditions for the harmonic emission. Finally, we present our conclusions in section 4.

2. Experimental

Theoretical and numerical method

Monolayer graphene-like materials are hexagonal crystal structures similar to graphene. Unlike graphene, one atom thickness 2D nanomaterial with planar structure, silicene and germanene have low-buckling structures, see as Figure 1. For this reason, they can possibly be analyzed with the tight-binding model. The silicene in our investigation is set up on substrate for aggregation protection. Thus, there is no aggregation in our consideration. Concerning spin dynamics induced by a rather large intrinsic spin-orbit coupling [2-8], they render as graphene-like materials topological insulators. Concerning their buckling structure, there is a layer separation between the two sublattices that makes the gap adjustable by applying a perpendicular electric field. As a result, the Hamiltonian for graphene-like materials has the following shape [2-8] that different graphene-like structures have different parameter values.

The silicene Hamiltonian is provided by the following expression:

$$\hat{H} = -t\sum_{\langle ij\rangle\alpha} \hat{c}^{\dagger}_{i\alpha}\hat{c}_{j\alpha} + it_{SO}\sum_{\langle ij\rangle\alpha} u_{ij}\hat{c}^{\dagger}_{i\alpha}(\sigma^{z})_{\alpha\beta}\hat{c}_{j\beta} - it_{\mathrm{int}R}\sum_{\langle\langle ij\rangle\rangle\alpha\beta} \mu_{ij}\hat{c}^{\dagger}_{i\alpha}(\vec{\sigma}\times\vec{d}_{ij})^{z}{}_{\alpha\beta}\hat{c}_{j\beta} + it_{\mathrm{ext}R}\sum_{\langle ij\rangle\alpha\beta} \mu_{ij}\hat{c}^{\dagger}_{i\alpha}(\vec{\sigma}\times\vec{d}_{ij})^{z}{}_{\alpha\beta}\hat{c}_{j\beta} \quad (1)$$

where $\hat{c}^{\dagger}_{i\alpha}$ creates an electron with spin polarization α and orbital state of $\Phi^{(1)}_{sp^3}$ at site i and $\hat{c}_{j\alpha}$ annihilates an electron with spin polarization α and π-orbital state $\Phi^{(1)}_{sp^3}$ at site j. Where, < i, j > run over all the nearest (next nearest) neighbor hopping sites. The first term represents nearest-neighbor hopping in which t is the hopping energy. The second term represents the effective spin-orbit coupling where - $\vec{u}_{ij} = (\vec{d}_i \times \vec{d}_j)/|\vec{d}_i \times \vec{d}_j|$ $\vec{d}_i$ and $\vec{d}_j$ denote the nearest bonds that connect the next nearest neighbors and $\vec{d}_{ij} = \vec{d}_i - \vec{d}_j$.

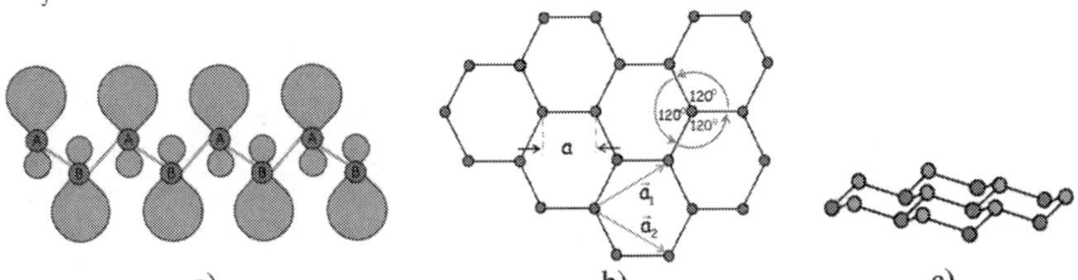

Figure 1. a) Schematic side view of the silicene π orbitals in buckled configuration. b) top and (c) side views of the crystal structure of the silicene monolayer.

The Hamiltonian is transformed into the k-space by using the relation $\hat{c}_{i\alpha} = \sum_i e^{-i\vec{k}\cdot\vec{R}_i}\hat{c}_{\vec{k}\alpha}$ in which $\vec{k} = (k_x, k_y)$ is the wave number of the charged carriers and $\vec{R}_i$ denotes the atomic positions. We also calculate the electronic structure of silicene. The energy dispersion surface clearly shows the Dirac cones at the Brillouin zone (BZ) boundary at the Fermi level (Figure 2a), which induces the massless Dirac fermion behavior near the Fermi level like graphene. The Dirac points are located in points K and K′ in the reciprocal space. K is related to K′ by inversion symmetry. The Γ point, at the center of the BZ, corresponds to the maximum gap; 0.172 eV for spin up of valence band to spin up of conduction band, 0.092 eV for spin down (up) of valence band to spin up (down) of conduction band, 0.014 eV for spin down of valence band to spin down of conduction band, as shown in Figure 2. For numerical calculations, we use available parameters for silicene and germanene reported in Refs. [8].

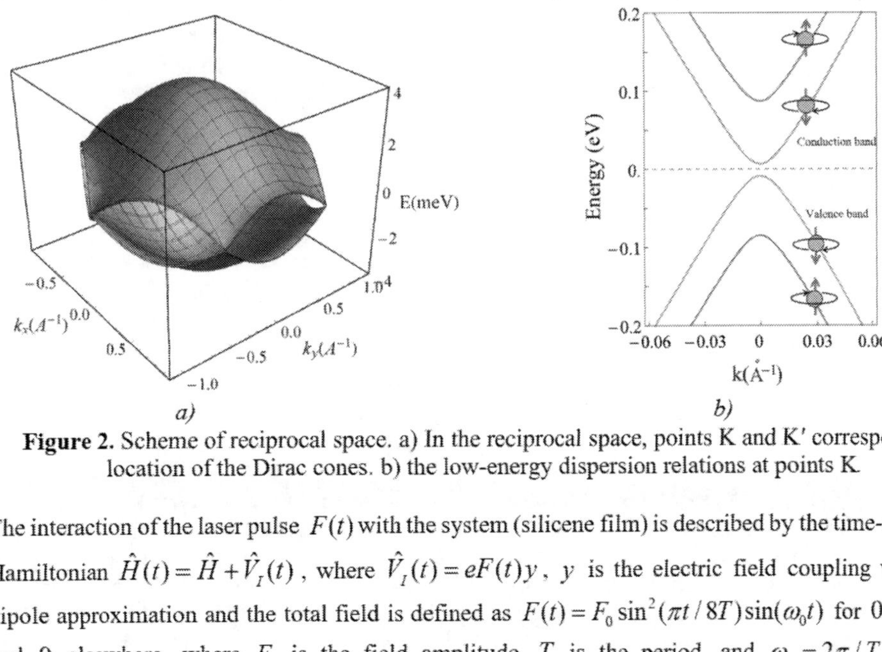

Figure 2. Scheme of reciprocal space. a) In the reciprocal space, points K and K′ correspond to the location of the Dirac cones. b) the low-energy dispersion relations at points K.

The interaction of the laser pulse $F(t)$ with the system (silicene film) is described by the time-dependent Hamiltonian $\hat{H}(t) = \hat{H} + \hat{V}_I(t)$, where $\hat{V}_I(t) = eF(t)y$, y is the electric field coupling within the dipole approximation and the total field is defined as $F(t) = F_0 \sin^2(\pi t/8T)\sin(\omega_0 t)$ for $0 \le t \le 8T$, and 0 elsewhere, where F_0 is the field amplitude, T is the period, and $\omega_0 = 2\pi/T$ the field frequency. We use the numerical method to solve the time-dependent Schrodinger equation

IOP Conf. Series: Materials Science and Engineering 1234 (2022) 012033 doi:10.1088/1757-899X/1234/1/012033

$\hat{H}(t)|\Psi(t)\rangle = i\hbar\partial/\partial t|\Psi(t)\rangle$. According to Larmor's semiclassical formula, the harmonic emission is given by the dipole acceleration. The mean value of the dipole can be written as

$$\vec{d}(t) = \langle\Psi(t)|e\vec{r}\,|\Psi(t)\rangle \tag{2}$$

and the HHG spectrum corresponding to Fourier transformation of $\vec{d}(t)$.

3. Results and discussion

HHG from silicene

Considering HHG spectrum from electron transition in each figure, we discover that the even harmonics in every figure are relatively low. According to the work [9], the inversion symmetry in monolayer graphene-like materials diminishes all of the even harmonics. As a result, only odd harmonics are significant in silicene. From all of the simulations, we find that the HHG in Figure 4.a is highest. This figure is the HHG of the electron transition from "the spin down valence band" to "the spin down of the conduction band."

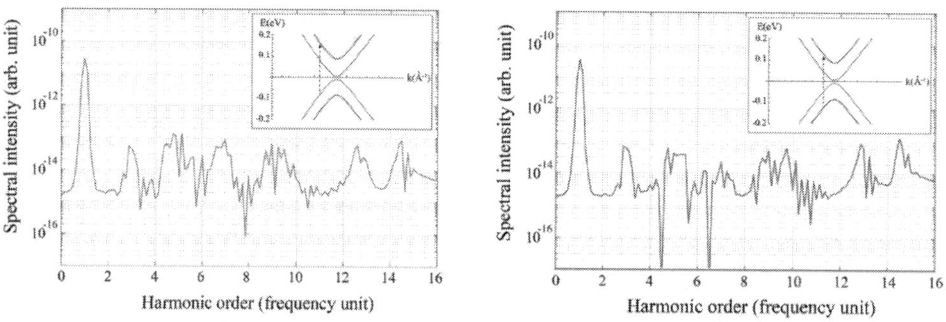

a) b)

Figure 3. Scheme of HHG spectrum corresponding to the transition process as the electron transition from a) the spin down valence band and b) the spin up valence band to spin up of conduction band, respectively.

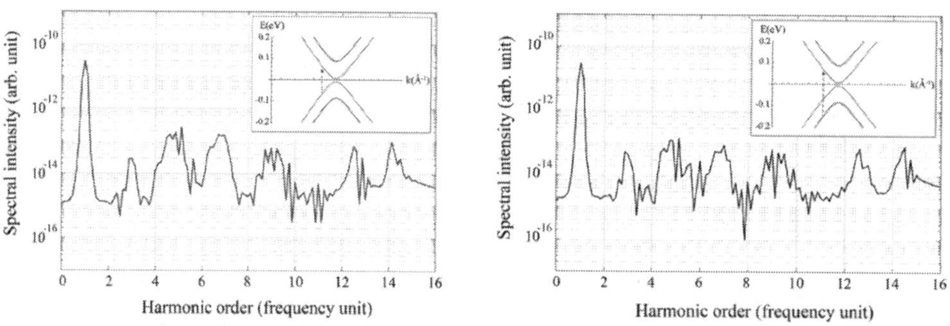

a) b)

Figure 4. Scheme of HHG spectrum corresponding to the transition process as the electron transition from a) the spin down valence band and b) the spin up valence band to spin down of conduction band, respectively.

The energy band gap difference of this transition is relatively lowest. On the contrary, the HHG in Figure 3.b is lowest. This lowest HHG is derived from the electron transition from "the spin up valence band" to "spin up of conduction band" that has the highest energy band gap difference relatively. We can infer, according to all simulations, that the less the different energy (between conduction band and valence band), the more the HHG. Furthermore, investigating Figure 3.b scrupulously, we see that the spectrum for 5th and 7th harmonic are rather low related to the other figures. That is to say, other than the Figure 3.b, the odd HHG in other figures remains fully. In addition, comparing with the work [10], we discover that the spectrum in Figure 4.a is rather similar to the graphene's. On this ground, the lower energy band gap difference makes the HHG in common with the graphene's. Accordingly, if the energy band gap difference in silicene increases, then the HHG in silicene will be more different than graphene's.

4. Conclusions

In summary, we have investigated high-harmonic emission from monolayer silicene driven by in-plane electric fields using first-principles TDSE calculations. By the silicene π orbitals that are not in the same plane, the spin orbit coupling, thereby, is significant. The graphene, on the other hand, has the spin orbit coupling as well but relatively insignificant. The spin orbit coupling is the essential factor that makes low energy band structure separating when the laser pulse field interacts with the material. Since there are the electric dipoles inside silicene, the electric field from the laser pulse certainly perturbs the dipoles. Thus, if the laser pulse field is oscillating, then the dipoles in silicene will vibrate according to Larmor's semiclassical formula. Such vibrating dipoles create emission photons that have odd high harmonics. The even harmonics, however, diminish owing to the inversion symmetry in silicene. The HHG from silicene has different characteristics depending on the electron transition. This transition can originate from "the spin down valence band" and "the spin up valence band" to "spin down of conduction band" and "spin up of conduction band", thereby being four different types of transition. From our simulations, the less the energy band gap difference of the electron translation, the more the HHG. Additionally, HHG of the electron transition (from the lowest energy band difference) rather resembles the HHG from graphene. Our investigation, moreover, shows that the laser pulse field applying on silicene determines the characteristic of HHG. On this ground, our work can be applied in attosecond physics for adjusting HHG and providing another way of HHG.

Acknowledgments
We acknowledge helpful discussions with Philip (Haiyong) for discussions. This work was supported by Science Achievement Scholarship of Thailand (SAST), Development and Promotion of Science and Technology Talents Project (DPST), specialized center of rubber and polymer materials in agriculture and industry (RPM) and Faculty of Science Kasetsart University for partial support.

References

[1] Bai, Y., Fei, F., Wang, S. et al. High-harmonic generation from topological surface states. Nat. Phys. 17, 311–315 (2021).
[2] Fanbanrai, P., A. Hutem, and S. Boonchui, Physica B Condens. Matter 472, 84-90 (2015).
[3] S. Boonchui and P. Nualpijit, Opt. Express 27, 28350-28363 (2019).
[4] Thitapura, T., Liewrian, W., Jutarosaga, T. et al., Plasmonics 12, 1439-14475 (2017).
[5] Thiti Thitapura, Watchara Liewrian, Tula Jutarosaga, and S. Boonchui, Opt. Express 25, 25588-25601 (2017).
[6] Ongart Suntijitrungruang, C. Summueang, and S. Boonchui, Opt. Mater. Express 11, 2854-2868 (2021).
[7] B. Lalmi, H. Oughaddou, H. Enriquez, A. Kara, S. Vizzini, B. Ealet, and B. Aufray, Appl. Phys. Lett. 97, 223109 (2010).
[8] Huon Shahabi, N., Phirouznia, Sci Rep 10, 16612 (2020).
[9] Shcherbakov, M.R., Zhang, H., Tripepi, M. et al. Nat Commun 12, 4185 (2021).
[10] Óscar Zurrón, Antonio Picón and Luis Plaja, New J. Phys. 20 053033 (2018).

IOP Conf. Series: Materials Science and Engineering 1234 (2022) 012034 doi:10.1088/1757-899X/1234/1/012034

Investigation of freshly prepared AgCl for high yield silver nanowires under polyol method

S Yala[1] and N Petchsang[1,2]*

[1]Department of Materials Science, Faculty of Science, Kasetsart University, Bangkok 10900, Thailand
[2]Specialzed Center of Rubber and Polymer Materials for Agriculture and Industry (RPM), Faculty of Science, Kasetsart University, Bangkok 10900, Thailand

*Corresponding author: nattasamon.p@ku.ac.th

Abstract. Silver nanowires (Ag NWs) have attracted much interest in academics and industries because of their potential applications in electronic devices such as flexible displays, pressure sensors, temperature sensors, and conductive layers. Generally, silver nanowire synthesized by the polyol method is the most common and promising chemical reaction. However, its by-products, a mixture of Ag NWs and silver nanoparticles, reduced Ag NWs performance. In this work, we reported the chemical method that is a fast and straightforward synthesis of Ag NWs by using freshly prepared silver chloride to avoid its degradation. The result showed that the freshly prepared silver chloride plays an essential role in promoting the formation of Ag NWs. The by-product was lower compared to using commercial silver chloride. We also found that avoiding boiling bubbles in flask via controlling temperature and reaction time by injecting ethylene glycol, resulting in decreased nanowires degradation. Furthermore, instantaneously decreasing temperature after finishing the cooking time helps reduce its by-product. A high yield of Ag NWs is noticeable under the observed synthetics.

1. Introduction

Silver nanowires (Ag NWs) are an inexpensive and flexible component of transparent conductive layers with a wide range of applications, including nanoelectronics with low optical extinction [1, 2], temperature sensors [3], strain and pressure sensors [4], substrates for surface-enhanced Raman [5]. Silver nanowires are commonly made using the polyol synthesis method, which involves dissolving silver nitrate in ethylene glycol with PVP as a stabilizing agent and a metal halide [6]. A wet chemical technique is the most promising method for generating Ag NWs. On the other hand, by-products might occur during the process, lowering Ag NWs performance. Although this technique has several advantages, it frequently produces a mixture of Ag NWs and silver nanoparticles as by-products, which cannot be ignored [7]. This study aims to use freshly prepared silver chloride to reduce the development of by-products during the synthesis of Ag NWs. We also investigated the different amounts of freshly prepared silver chloride and the injection rate of ethylene glycol to maintain the constant temperature. These two factors are crucial in improving Ag NWs while drastically reducing their by-products.

2. Experimental

Chemicals and Materials. Commercial Silver chloride (AgCl, 99.9%) and silver nitrate ($AgNO_3$,99.9+%) were purchased from Alfa Aesar. Sodium chloride (NaCl ,99+%) was purchased from Acros Organics. Polyvinylpyrrolidone (PVP, Mw ≈ 55000, powder) and ethylene glycol (anhydrous, 99.8%) were purchased from Sigma- Aldrich. Acetone (99.5%) was purchased form LOBA Chemical and iso-propyl alcohol (IPA, ≥99.8%) was purchased from Fisher Scientific. All the chemicals were used without any purification step.

Synthesis of AgCl. This synthesis is performed in the dark due to the photosensitivity of AgCl. Silver nitrate aqueous solution (1 mL, 0.5 M) is mixed with sodium chloride aqueous solution (1 mL, 0.5 M) then silver chloride immediately precipitated. The silver chloride was clean by DI water several time and then store in acetone before using.

Synthesis of Ag NWs. A modified polyol method was used to synthesized Ag NWs. In briefly, 0.217 g of PVP (Mw ≈ 55000) was added to 60 mL of ethylene glycol in a 100 mL three-neck round-bottom flask. The solution was heated to 168 °C under stirring at 510 rpm. At the same time, $AgNO_3$ 0.3 g was dissolved by EG 0.6 mL in a separated vial. Once the solution has a stable temperature, the freshy prepared AgCl was added and then the solution turned from transparent to light yellow. After 20 min, the prepared $AgNO_3$ solution was injected to the flask. Ag NWs appeared as gray color and was cooked for a designed time before quickly cooling the solution to room temperature.

Measurements. The UV-VIS Spectrophotometer (UV-vis) and Scanning electron microscopy (SEM).

3. Results and discussion

Freshly synthesized AgCl was used instead of commercial AgCl to create Ag NWs. Figure 1 shows SEM images of Ag NWs created with commercial AgCl and freshly synthesized AgCl. Figure 1a showed a mixture of nanowires and nanoparticles of various sizes and shapes when Ag NWs were made by commercial AgCl. In contrast, the by-products of Ag NWs created with freshly synthesized AgCl (Figure 1b) are barely visible.

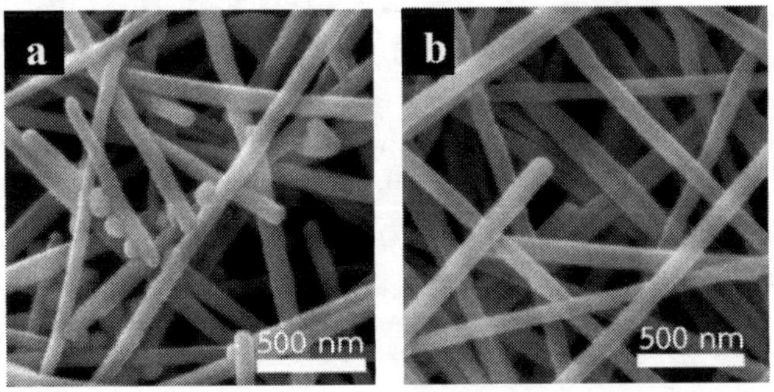

Figure 1. Scanning electron microscopy images of Ag NWs comparison by using (a) commercial AgCl and (b) freshly prepared AgCl.

Moreover, UV–vis spectra of Ag NWs made commercial AgCl and freshy prepared AgCl are conducted as seen in Figure 2. The results show the peak of Ag NWs using freshly generated AgCl is narrower than the peak of Ag NWs using commercial AgCl. This indicated that Ag NWs using freshly prepared AgCl have a low yield of by-products than Ag NWs using commercial AgCl.

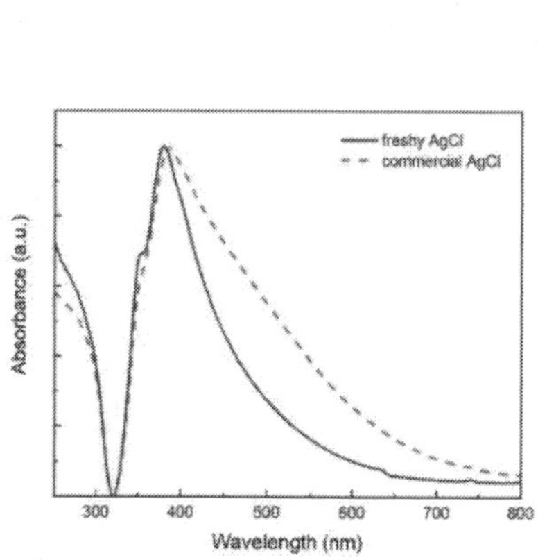

Figure 2. Absorbance of UV vis spectra of commercial AgCl (dash line) and freshy prepared AgCl (solid line).

Although freshly prepared AgCl can reduce nanoparticles, AgCl concentration is also crucial for Ag NWs formation. The amount of freshly prepared AgCl was investigated, including 1.7 mg, 10 mg, 15 mg and 20 mg. Figure 3 shows the SEM images of Ag NWs prepared at different concentrations of AgCl. Numerous nanoparticles are formed as by-products when 20 mg of AgCl is used (Figure 3d). The lowest concentration of AgCl 1.7 mg (Figure 3a), on the other hand, barely observed the by-products. The results are also confirmed by UV-vis spectra in Figure 4 as comparative spectra for various conditions. Figure 4a shows that Ag NWs synthesized with freshly prepared AgCl 1.7 mg has the narrowest peak compared to the others.

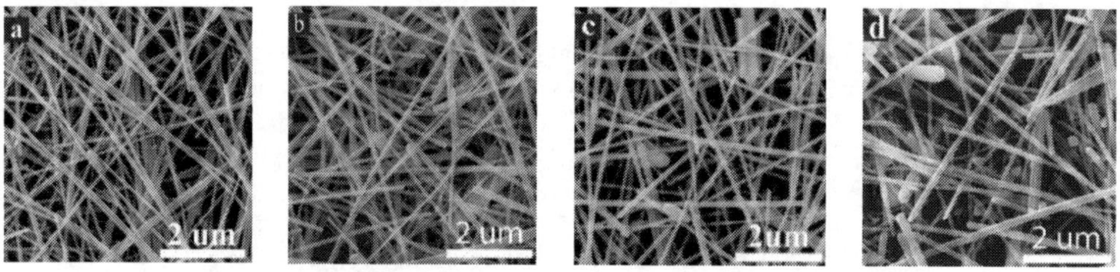

Figure 3. SEM images of Ag nanowires grown with the polyol process by various amounts of freshly prepared AgCl, which are 1.7 mg (a), 10 mg (b), 15 mg (c) and 20 mg (d).

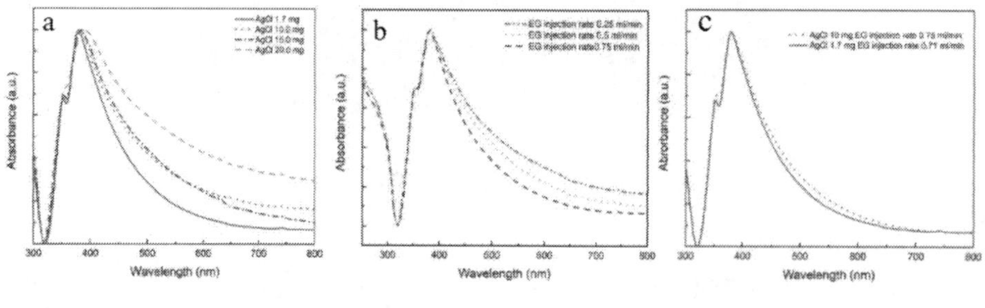

Figure 4. a. UV vis spectra of various freshy prepared AgCl concentration 1.7 mg (solid line), 10 mg (dot line), 15 mg (dash dot line) and 20 mg (dash line), b. UV vis spectra of varying EG injection rate 0.25 mL/min (dash dot dot line), 0.5 mL/min (dot line) and 0.75 mL/min (dash line), c. Comparison of UV vis spectra between AgCl 10 mg (dot line) and 1.7 mg (solid line) at the similar EG injection rate.

In addition to the AgCl study, we investigated injecting ethylene glycol to maintain a steady temperature in the process to avoid a boiling bubble. With the same amount of AgCl, multiple rates of injecting ethylene glycol are investigated, including 0.25 mL/min, 0.5 mL/min, and 0.75 mL/min. As seen in Figure 4b, the high-rate injection of ethylene glycol produces the lowest amount of by-products while the low-rate injection produces a higher amount of by-products. Therefore, the injection rate plays a role in minimizing the by-products of Ag NWs. We also compare the similar injection rate with different amounts of AgCl. Both UV curves are not much different at a similar injection rate around 0.75 mL/min with a high different amount of AgCl 1.7 mg and 10 mg (Figure 4c). By adjusting another parameter, we can compare the effects of the EG injection rate (Figure 4b) and the amount of AgCl (Figure 4a). Surprisingly, the data suggest that the rate of EG injection has a more significant impact than the amount of AgCl. It may also be demonstrated by setting the amount of AgCl and altering the EG injection rate (Figure 4b), resulting in UV curve broadening. Furthermore, immediately lowering the temperature after the cooking period helps reduce by-products.

4. Conclusions

In this experiment, we present a simple and fast one-pot method to synthesize a high yield of silver nanowires along with reducing its by-products. The result showed that the freshly prepared silver chloride plays an essential role in reducing by-products compared to commercial silver chloride. In addition, we also found that avoiding boiling bubbles in reaction through maintaining temperature by injecting ethylene glycol will lower the by-products.

Acknowledgement

This research is partially funded by Kasetsart University through the Graduate School Fellowship Program. We also thank to the department of Materials Science, Faculty of Science, Kasetsart University for the support.

References

[1] Zhang P, Wyman I, Hu J, Lin S, Zhong Z, Tu Y, Huang Z and Wei Y 2017 *Mater. Sci. Eng.* **1** 23.
[2] Langley D, Giusti G, Mayousse C, Celle C, Bellet D and Simonato J 2013 *Review. Nanotechnology* **452001** 421021.
[3] Lu Y, Wang M. T, Hao C. J, Zhao Z. Q and Yao J. Q 2014 *IEEE Photonics J.* **1.**
[4] Hu W, Niu X, Zhao R and Pei Q 2013 *Appl. Phys. Lett.* **083303**
[5] Tao A, Kim F, Hess C, Goldberger J, He R, Sun Y, Xia Y and Yang P 2003 *Nano Lett.* **1229** 1233.
[6] Da Silva R. R, Yang M, Choi S. Il, Chi M, Luo M, Zhang C, Li Z. Y, Camargo P. H. C, Ribeiro S. J. L and Xia Y 2016 *ACS Nano.* **7892** 7900.
[7] Parente M, van Helvert M, Hamans R. F, Verbroekken R, Sinha R, Bieberle-Hutter A and Baldi A 2020 *Nano Letters.* **5759** 5764.

The effect of shot peening on corrosion performance of anodized laser powder bed fusion manufactured AlSi10Mg

T Rautio*, A Hamada, J Kumpula and A Järvenpää
Kerttu Saalasti Institute, University of Oulu, Pajatie 5, 85500 Nivala, Finland

*Corresponding author: timo.rautio@oulu.fi

Abstract. Anodizing is commonly used on aluminium to improve the surface properties of the product for better corrosion resistance and wear resistance. In this research, the anodizing of laser powder bed fusion (LPBF) manufactured AlSi10Mg parts which were subjected to prior shot peening (SP) were investigated. Anodized specimens were analysed using SEM imaging and the corrosion properties were examined considering the effect of residual stress relief heat treatment and the SP process. The results showed that the SP deteriorates the corrosion performance of the material as such, but together with the subsequent anodizing it resulted to the lowest corrosion current of all investigated structures. As the SP can be used to lower the surface roughness of the LPBF parts and to increase fatigue life as well, the results of this work further encourage the use of SP with anodizing.

1. Introduction

AlSi10Mg is currently the most used aluminium alloy in additive manufacturing (AM) by laser powder bed fusion (LPBF). This material has several advantages including light weight, high strength, and good corrosion resistance [1]. Due to the difference in microstructure and defect structure, the LBPF parts have better corrosion properties compared to cast alloy [2]. Material properties can be further enhanced with suitable post processing methods or surface finishing. In authors' previous work silver shell-copper core coating could be used to increase the electrical conductivity and improved corrosion resistance [3]. In addition, the rough surface of LPBF manufactured part could be smoothed with the aid of shot peening (SP). Anodizing is commonly used to improve the surface qualities of aluminium parts. Previous studies have focused on the effect of heat treatment [4], and the effect of the characteristic microstructure of the LPBF material [5].

The effect of shot peening on the corrosion properties of LBPF manufactured AlSi10Mg has not been previously studied. The current work addresses this with a microscopical view of the anodized surface together with corrosion resistance measurements.

2. Experimental

2.1. Specimen manufacturing and the material

An SLM 280 HL printer was used for the LPBF manufacturing of the samples. The powder was supplied by Carpenter Additive and had a size range of 20-63 μm with a chemical composition detailed in [3]. The following parameters were used to print all of the samples in 90° orientation: power (P) of 650 W, laser speed (v) of 1850 mm/s, hatch spacing (h) of 0.17 mm, layer thickness (t) of 60 μm and laser spot diameter of 0.15 mm, which resulted to an energy density of 34.4 J/mm^3. Stress relief heat treatment (T5-HT) was carried out at 300°C for 2 h in a muffle furnace under argon atmosphere.

2.2. Anodizing and shot peening processes
Samples were anodized after the printing and the subsequent HT. The process consisted of several steps and had a rinse after each one. At first, the samples underwent an alkaline prewash for 15 minutes which was followed by activation in nitric acid for 10 min. Sulfuric acid anodizing was completed in 55 min and the process was finished by cold-sealing for 15 min and ageing for 15 min and drying. Shot peening was performed in a shot peening chamber using glass beads in the size of 0.25-0.5 mm and utilizing a shot pressure of 3 bar.

2.3. Characterization
AlSi10Mg alloy and its microstructure and surface features with and without the anodizing were analyzed using a field-emission gun scanning electron microscope (FEG-SEM) (Carl Zeiss Ultra plus). Secondary electron micrographs were captured at an operating voltage of 5 kV.

Corrosion resistance of the structures were evaluated in an Autolab electrochemical system (Autolab VersaSTAT 3 by Princeton Applied Research, USA). All the electrochemical experiments were performed in 3.5% NaCl solution at room temperature using a three-electrode compartment glass cell. Round samples were cut with a 15 mm diameter and worked as electrodes with an exposed area of 10 mm^2. The potentiodynamic polarization curves were recorded during immersion for 30 min until a stable open circuit potential (OCP) was reached.

3. Results and discussion
3.1. Surface features and microstructure of the LPBF AlSi10Mg
The surface features of the LPBF manufactured AlSi10Mg are presented in Figure 1a). The layer- by-layer formation of the part from bottom to top can be identified from the surface features that appear approximately at the build layer thickness intervals of 60 μm. The surface has a rough appearance due to the partially melted powder particles that are still attached to the surface. In closer inspection the surface has small cracks and pits that are mostly related to the areas between the layers. Cross-section of the AlSi10Mg after the HT is presented in Figure 1b). The microstructure of the material is pseudo-eutectic with primary α-Al phase (areas in size range of 1-2 μm) surrounded by precipitated Si-network.

3.2. Anodized AlSi10Mg
Cross-sectional view of the anodized AlSi10Mg is presented in Figure 2a. A very thin platinum coating was developed over the sample to be able identify the layer thickness of the non- conducting anodized layer. The anodized layer is visible in Figure 2a and has a thickness of 3-4 μm. Figure 2b shows how the anodizing has thoroughly penetrated into the material voids and cracks. Although the surface features have very abrupt changes and discontinuities, the anodizing shows very uniform around each of those.

3.3. Surface features and microstructure of the LPBF AlSi10Mg
The corrosion resistance of the LPBF manufactured AlSi10Mg material in as built, HT and HT + SP conditions were measured and compared to the respective anodized structures. The results (Figure 3 and Table 1) show that heat treatment has a beneficial effect on the corrosion resistance, but the usage of SP reduces it (i_{corr} raises to 50 μA/cm^2 from 13). Anodizing has a clear positive impact on the corrosion properties on all investigated structures. The lowest corrosion current density was measured with the combination of HT and SP before anodizing. It was determined that the SP closing the surface voids and cracks and removing the balling phenomenon and satellites as shown in [3] is beneficial for the uniformity of the anodized layer improving the corrosion resistance as well. This is due to the decreased surface porosity, which is an important corrosion parameter. Reduced surface roughness reduces the

pitting corrosion as there are fewer and smaller sites for pit initiation, but the residual compressive stresses introduced by SP have detrimental effect but can be suppressed by anodizing.

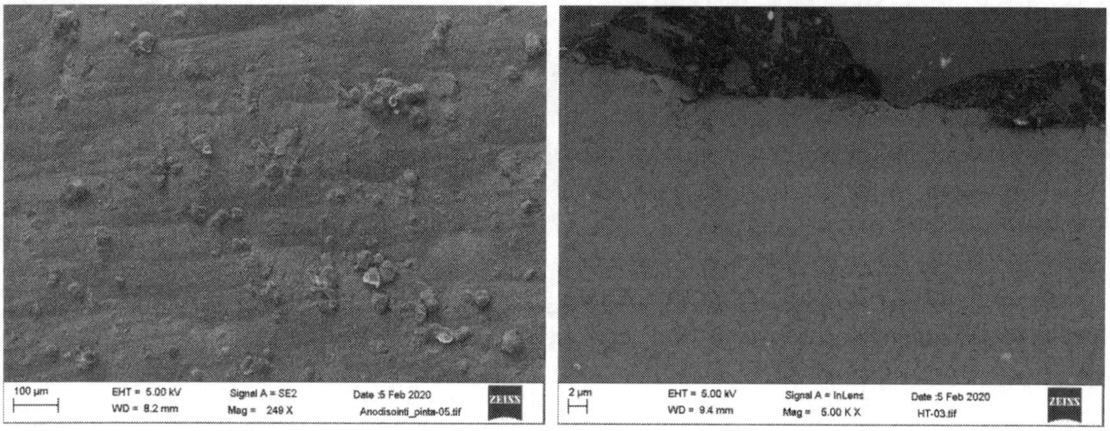

Figure 1. a) Surface features of the laser powder bed fusion manufactured AlSi10Mg and b) cross-section showing the microstructure.

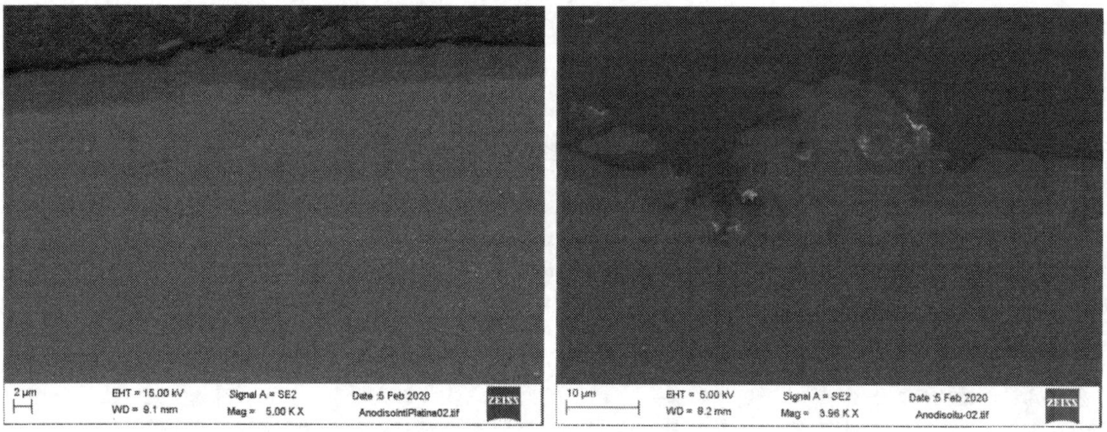

Figure 2. a) Anodized layer on laser powder bed fusion manufactured AlSi10Mg smooth area and b) around rough areas.

4. Conclusions

This work investigated the effect of shot peening on the properties of laser powder bed fusion (LPBF) manufactured AlSi10Mg and can be concluded as follows:

- The surface of LPBF is rough due to the partially melted particles.
- Anodized layer was successfully built on all investigated structures and resulted to improved corrosion resistance in all cases.
- The highest corrosion resistance was measured with the combination of heat treatment and SP before the anodizing.

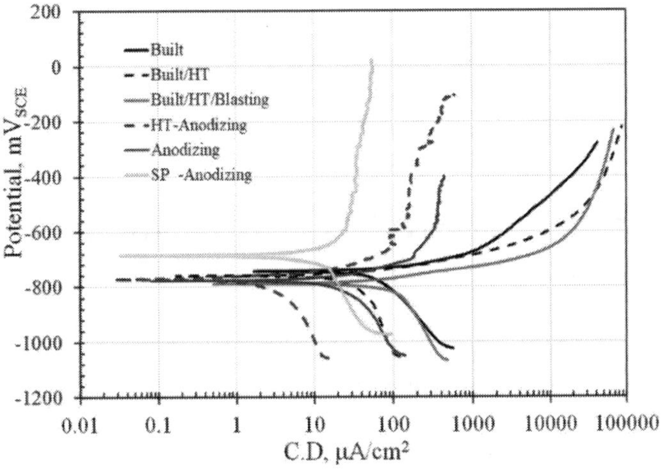

Figure 3. Potentiodynamic polarization curves of additively manufactured AlSi10Mg in different states.

Table 1. Corrosion current density i_{corr} and corrosion potential E_{corr} of different states of additively manufactured AlSi10Mg in 3.5% NaCl solution.

State	**E_{corr} (mV$_{SCE}$)**	**i_{corr}(μA/cm^2)**
Built	-740	30
HT	-760	13
HT+SP	-788	50
Anodizing	-778	10
HT + Anodizing	-772	4
HT + SP + Anodizing	-685	3

Acknowledgement
The authors would like to acknowledge the financial support received from the Council of Oulu Region and the European Union (European Regional Development Fund) for the "Hybridi" and M3D projects.

References

[1] Dresler N., Inberg A., Ashkenazi D., Shacham-Diamand Y. and Stern A. 2019 Metallography, Microstructure, and Analysis, **8(5)** 678–692
[2] Leon A., Shirizly A. and Aghion E.. 2016 Metals, **6(7)** 148
[3] Rautio T., Hamada A., Kumpula J., Järvenpää A. and Allam T. 2020 Surface and Coatings Technology, **403** 126426
[4] Rubben T., Revilla R.I. and De Graeve I. 2019 Journal of The Electrochemical Society, **166(2)** C42–C48
[5] Revilla R.I., Terryn H. and De Graeve I. 2018 Journal of The Electrochemical Society, **165(9)** C532–C541

IOP Conf. Series: Materials Science and Engineering 1234 (2022) 012036 doi:10.1088/1757-899X/1234/1/012036

Investigation of heated silver nanowires under surface modification

T Khamla[1] and N Petchsang[1, 2*]

[1] Department of Materials Science, Faculty of Science, Kasetsart University, Bangkok 10900, Thailand
[2] Specialized Center of Rubber and Polymer Materials for Agriculture and Industry (RPM), Faculty of Science, Kasetsart University, Bangkok 10900, Thailand

*Corresponding author: nattasamon.p@ku.ac.th

Abstract. Silver nanowires (Ag NWs) have received much attention and applied in many fields due to their excellent conductivity. Hence, Ag NWs were usually integrated with low conductive materials to enhance the conductivity of the overall system. In some conditions, the surface of Ag NWs need to be modified before combining with other materials. There are many ways to change and prepare the surface of Ag NWs, for example, ligand exchange which is a simple process. TOPO (trioctylphosphine oxide) is a robust coordinating ligand and is classified as an essential component for low-dimensional nanomaterials. In comparison, ODE (1-octadecene) is a noncoordinating ligand that is often used in quantum nanomaterials preparation. In this experiment, we investigated the surface attachment for ligand exchange of TOPO and ODE on the surface of Ag NWs under variations of reaction times at 150°C. The scanning electron microscope showed the rough surface when TOPO was added, and there was no noticeable change of surface when ODE was added. These results are promising for the preparation of heterostructure in the future.

1. Introduction

The study of one-dimensional (1D) material nanowire formations has gotten much attention because the integration of components impacts their performance in terms of electrical, optical, and thermal qualities. Silver (Ag) nanowires, in particular, were always chosen as the initial part for further combination with other materials such as ZnO, SiO_2, TiO_2, SnO_2, Fe_3O_4 due to their highly conductive and transparent qualities. [1-6] As a result, the surface modification with ligand exchange technique was used to prepare Ag nanowires for use. A ligand is an essential component in this process, especially strong coordinating ligands like TOPO (trioctylphosphine oxide) and weak coordinating ligands as ODE (1-octadecene). Both ligands were expected to be appropriate solvents for synthesizing low-dimensional materials. [7] In this experiment, we investigated the attached ligand of either TOPO or ODE and both of them on the surface of Ag NWs under variations of reaction times at temperature 150°C. The thickness and structure of the material were examined using a scanning electron microscope (SEM) and X-ray diffraction pattern (XRD). This method can be further used to create heterostructure for another application.

2. Experimental

2.1 Chemicals

Silver nanowire solution (4%W/W in IPA). Trioctylphosphine oxide (TOPO, 99%) was purchased from Sigma-Aldrich and 1-Octadecene (ODE) (Technical grade 90%) was purchased from Alfa-Aesar, Toluene and Ethanol were purchased from Fisher. All of these were used without any further purification.

2.2 Experiment

For ligand attached on Ag NWs surface, TOPO 1.0 g (2.586 mmol) and ODE 15 mL were added to a round bottom flask under the presence of 0.3 mL of the as-prepared 4% W/W Ag NWs solution. The mixed solution was stirred under 620 rpm and heated to 100°C to remove impurities and isopropanol from the Ag NWs solution. After 30 min, the solution temperature was increased to 150°C under nitrogen gas. When the temperature reached this point, the reaction was cooked for 30 minutes. After that, the solution was cooldown by immersing in a water bath for 5 min, and then it was centrifuged at 4,400 rpm followed by two times washing with toluene and ethanol to remove excess TOPO and storing in toluene.

3. Results and discussion

As-prepared Ag nanowires synthesized by polyol method with smooth surface and uniform were characterized by scanning electron microscope (SEM) as shown in figure 1. Diameter of NWs approximately 90 $\pm$ 8 nm with 8.7 μm in length.

Figure 1. SEM images of as-prepared silver nanowires with a) 20kx and b) 60kx magnification.

Following that, Ag nanowires surface were modified with either TOPO or ODE solvent at 150°C for 5 min. Figure 2a depicts the rough surfaces of TOPO, whereas Figure 2b depicts the smooth surfaces of ODE solvent. The average diameter of modified Ag NWs in TOPO is 88.70 nm, whereas it is 87.12 nm in ODE. It means that TOPO can attach to the surface of Ag nanowires and no ODE on the surface of Ag nanowires.

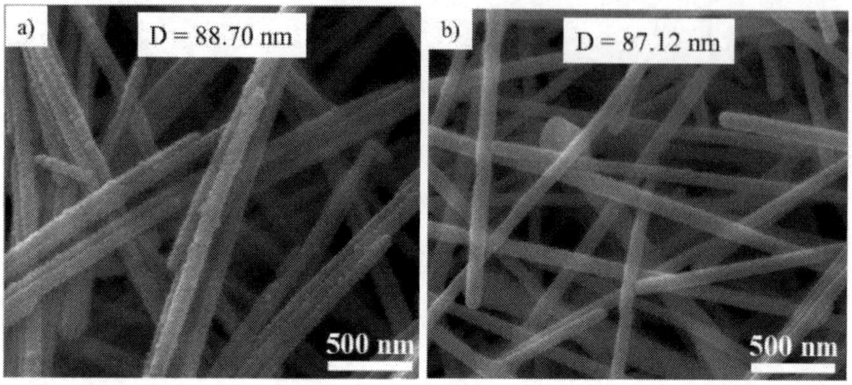

Figure 2. SEM images and diameter of modified Ag nanowires surface under 150°C with: a) TOPO and b) ODE solvent.

Under a combined TOPO and ODE solvent with a cooking temperature of 150°C for 5 min, the rough surface from TOPO is also apparent. The result is obtained with a larger diameter at 91.21 nm (Figure 3a). When the reaction time increased to 30 min, the rough surface still appeared, and the diameter of modified Ag NWs surface gradually increased to 93.76 nm, as seen in Figure 3b. The increasing diameter depicted the thickness of modified Ag NWs as a function of reaction time.

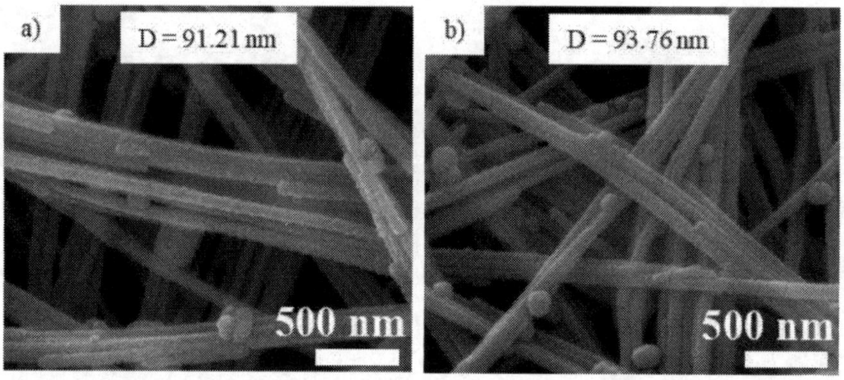

Figure 3. SEM images and diameter of Ag nanowires with co-operating TOPO and ODE solvent under 150°C with reaction time: a) 5 and b) 30 min.

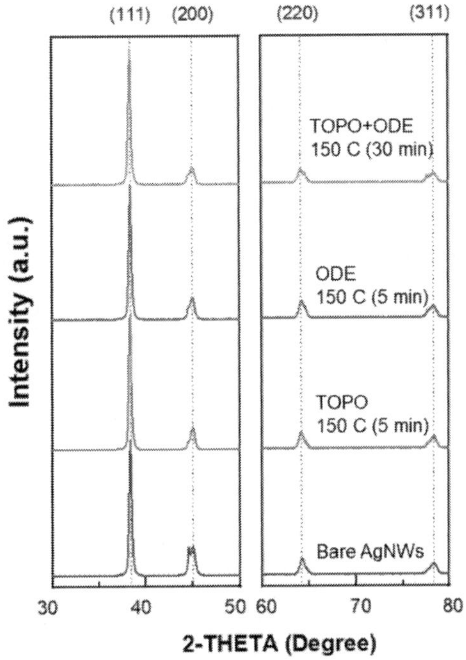

Figure 4. XRD patterns of silver nanowires with either TOPO or ODE and both at 150°C with reaction time 5 and 30 min.

Finally, the XRD pattern was used to analyze the surface-modified Ag NWs structure. The XRD pattern exhibits peaks of bare Ag NWs and surface-modified Ag NWs at different conditions, as shown in Figure 4. The four peak positions correspond to the FCC structure: (111), (200), (220), and (311). The XRD pattern revealed that the structure of Ag NWs was not damaged and unchanged during the modification process.

4. Conclusions

The surface modification of Ag nanowires in our experiment proposed ligand exchange with TOPO or ODE under 150°C. TOPO can attach to the surface of Ag nanowire, while ODE has no effect. Furthermore, when TOPO and ODE are mixed with increasing the reaction time, the results show a thicker rough surface in SEM image with gradually increasing diameter without deformation of Ag nanowires structure. Therefore, these results are promising for the preparation of integration of Ag nanowires with other materials in the future.

Acknowledgement

This research is partially funded by Kasetsart University through the Graduate School Fellowship Program. This research was supported in part by the National Nanotechnology Center (NANOTEC), NSTDA, the Ministry of Science and Technology of Thailand, through its program of Research Network of NANOTEC (RNN)

References

[1] Kim KP, Hwang DK, Kim DH and Woo SH 2017 *J. Nanosci. Nanotechnol.* **17** 8201.
[2] Wang Y, Zhou X, Liang C, Li P, Hu X, Cai Q, Zhang Y, Li F, Li M and Song Y 2017 *Adv. Electron. Mater.* **3** 1700169.
[3] Huang Y, Tian Y, Hang C, Liu Y, Wang S, Qi M, Zhang H and Peng Q 2019 *ACS Appl. Nano Mater.* **2** 1738.

[4] Baranowska-Korczyc A, Mackiewicz E, Ranoszek-Soliwoda K, Nejman A, Trasobares S, Grobelny J, Cieslak M and Celichowski G 2021 *RSC Adv.* **7** 4174.
[5] Ma J, Wang K and Zhan M 2015 *ACS Appl. Mater. Interfaces.* **7** 16027.1
[6] Opoku F, Govender KK, Catharina Elizabeth van Sittert CG, Govender PP 2017 *Adv. Sustain. Syst.* **1** 1700006.
[7] Dhaene E, Billet J, Bennett E, Van Driessche I and De Roo J 2019 *Nano Lett.* **19** 7411.

IOP Conf. Series: Materials Science and Engineering 1234 (2022) 012037 doi:10.1088/1757-899X/1234/1/012037

Application of near infrared spectroscopy for quality evaluation in Para rubber industry

D Siano[1], A Puttipipatkajorn[2] , A Puttipipatkajorn[3], S Noypitak[1] and A Terdwongworakul[1,*]

[1] Department of Agricultural Engineering, Faculty of Engineering at Kamphaeng Saen, Kasetsart University, Nakhon Pathom 73140, THAILAND.

[2] Department of Food Engineering, Faculty of Engineering at Kamphaeng Saen, Kasetsart University, Nakhon Pathom 73140, THAILAND.

[3] Department of Computer Engineering, Faculty of Engineering at Kamphaeng Saen, Kasetsart University, Nakhon Pathom 73140, THAILAND.

* Corresponding author:: fengant@ku.ac.th

Abstract. Quality of products manufactured from Para rubber was an important aspect for merchandizing. Near infrared spectroscopy in a wavelength range of 900 to 1700 nm, which was a rapid and non-destructive technique, was applied to create quality predicting models for various Para rubber products. Near infrared hyperspectral imaging (NIR-HIS) technique was implemented to predict dry rubber content (DRC) of cup lump. Classification of adulterated Para rubber sheet was investigated using NIR-HSI. And NIR portable system was also developed for moisture content determination of Para rubber timber. The results showed that the DRC of the cup lump could be predicted with high accuracy with the coefficient of determination (R^2) of 0.99 and root mean square of error of prediction (RMSEP) of 0.64%. Regarding the Para rubber sheet, the sulphuric acid coagulated sheets was classified from the formic acid coagulated sheet very good accuracy of 98.3%. Finally, The NIR portable system was developed and could be used to determine the moisture content of the timber with R^2 and RMSEP of 0.96 and 5.37%, respectively. The application of NIRS was successful and provided feasibility for development into a future commercial use.

1. Introduction

Para rubber is one of the strategic crops in Thailand as it generates very high income and has promoted Thailand to one of the biggest exporters of Para rubber. At present, farmers commonly make rubber cup lumps from latex to add value to the latex. The cup lump can be further used to produce rubber blocks. Merchandizing the Para cup lump is based on the dry rubber content (DRC) in the cup lump. However, the buyer evaluates the DRC by visual inspection which is a subjective method and depends on the experiences of the buyer. Rubber sheet is one form of early processing from the latex. The popularity of this form is mainly attributable to the low cost and simplicity of the processing machinery. One common problem in the rubber sheet processing is the rubber sheet adulteration, which occurs during the coagulation process by the addition of acids other than formic acid. The common methods for determining the type of acids used in the Para rubber sheet are based on parameters such as Mooney viscosity or the plasticity retention index. However, these common methods are time-consuming and destructive. Para rubber wood is product from Hevea trees that give low content of latex. The low latex producing trees are normally fell and transformed to sawn timber for exporting. The sawn timber from these trees have made a lot of profit to Thailand. Before exporting, sawn timber Hevea wood must be checked to ensure that the moisture content is lower than 12% dry basis following Thai Industrial Standards Institute.The sawn timber companies generally use commercially available pin moisture meter and digital moisture meter to determine the moisture content. However, the pin meter can only be used to measure the moisture content at a specific point and makes hole damage in the wood. Regarding the digital moisture meter, specific gravity of wood must be known to program in meter. In order to maintain the quality of the products from the Para rubber, a rapid and nondestructive testing technique is needed to retain standards.

NIR spectroscopy is a type of high-energy vibrational spectroscopy performed in the wavelength range 750-2500 nm. NIRS is a fast and nondestructive analytical method. The NIR light interacts with the molecules of C-H, O-H and N-H causing the bond to vibrate and absorbance occurrence. Therefore, the chemical content in the organic materials is sensitive to the NIR light. The absorbance of the NIR light is related to the quantity of the chemical content according to Beer-Lambert law. It has proven its effectiveness for both qualitative and quantitative analyses in several fields. Detection limits are typically 0.1%. Hyperspectral imaging has been integrated with NIRS which provides spatial information other than spectral information of the sample. In addition, more spectral information called hypercube is acquired by NIR-HSI. One important benefit of the NIR_HSI is that the color-mapped image of the chemical can be produced from the NIR-HSI image displaying the distribution of the chemical content across the sample surface.

Our research group has investigated the application of NIRS to evaluate the dry matter content (DRC) in the cup lump. The classification of adulterated rubber sheet was also achieved by the use of NIR-HSI. Finally, the moisture content determining system based on a portable spectrometer was developed in order to replace currently used devices which were either destructive or could be used only for a limited range of moisture content.

2. Experimental

2.1 Evaluation of dry rubber content of cup lump

This research developed a predictive model for the DRC of cup lump using NIR-HSI system. The result led to a development of a measuring system which could accurately evaluate the DRC in the cup lump which would help improve the quality control in the cup lump merchandizing.

The cup lump rubber was prepared from fresh latex collected in eastern Thailand in October 2017. The cup lump rubber was subjected to NIR-HSI scanning (Figure 1.) (Specim, Spectral Imaging Ltd., Finland). Three 20 W light sources shone the light on to the surface of the cup lump rubber placed on a moving translation stage at a speed of 10 mm/s. The calibration model for the DRC prediction was built by the least square support vector machine (LS-SVM)

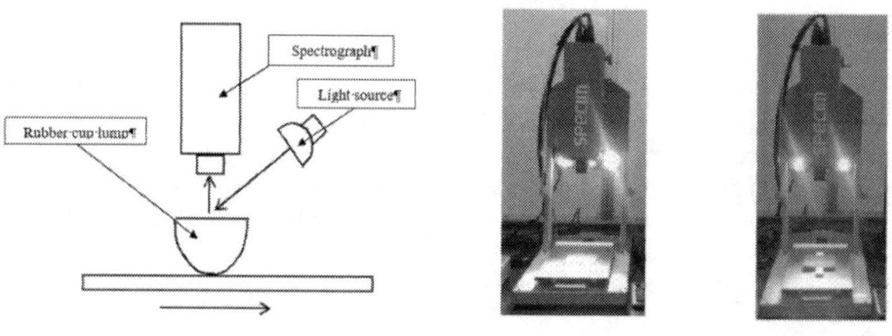

(a) (b)

Figure 1. (a) A cup lump rubber in scanned position by hyperspectral imaging system and (b) acquisition of the Para rubber sheet using the NIR-HSI system.

IOP Conf. Series: Materials Science and Engineering 1234 (2022) 012037 doi:10.1088/1757-899X/1234/1/012037

2.2 Classification of adulterated Para rubber sheet

The classification model was created using NIR-HSI system to separate adulterated Para rubber sheet. The result provided a means for quality control as the adulterated Para rubber sheet could not be visually differentiated from the normal sheet. The transflectance mode was achieved by placing the sheet on the translation stage and a spectralon with high reflective characteristic was placed underneath the sheet during the scanning to enable double light path through the sheet providing higher reflection back to the spectrograph. Partial least squares-discriminant analysis (PLS-DA) was performed to develop a classifying model with a class variable consisting of two levels as a response.

2.3 Development of NIR portable system for quality evaluation of Para rubber timber

This research developed a moisture content determining system for sawn timber, which comprised a portable spectrometer and a controlling software. The calibration model was first created and validated and then used as part of codes in controlling Android application. The housing of the spectrometer was designed and made to provide a hand-held system for measuring the moisture content in the timber which was used with a smartphone. The outcome of this research provided a system which could non-destructively measure the moisture content in timber. The developed system could replace a currently used device which had some limitation in the application.

Six hundred Para rubber sawn timber samples consisting of 5 different grades and three groups of moisture content were collected from sawmill companies in Southern, Eastern and Northeast regions of Thailand in 2019. Each sample was scanned twice at four positions by a portable spectrometer (DLP NIRscan Nano, Texas Instruments, Dallas, Texas, USA) in a range 900 nm to 1700 nm at an interval of 3.51 nm. The predictive model was developed using Unscrambler V.9.8 (Camo, Oslo, Norway) assigning the moisture content as the dependent variable and the NIR absorbance values for the whole range of wavelength as an independent variables.

3. Results and discussion

3.1 Evaluation of dry rubber content of cup lump

The reflectance spectra obtained from the NIR-HSI system showed prominent peaks (Figure 2.). At around 1200 nm, the second overtone vibration partially corresponded with C-H bonding of polymer [1]. The developed models were used to predict the DRC of all samples in the prediction set and the results are shown in Table 1. The model developed from spectral data of NIR-HSI system yielded very good performance providing the residual predictive deviation of calibration (RPD) greater than 5.0, which was considered good for application of quality control [2].

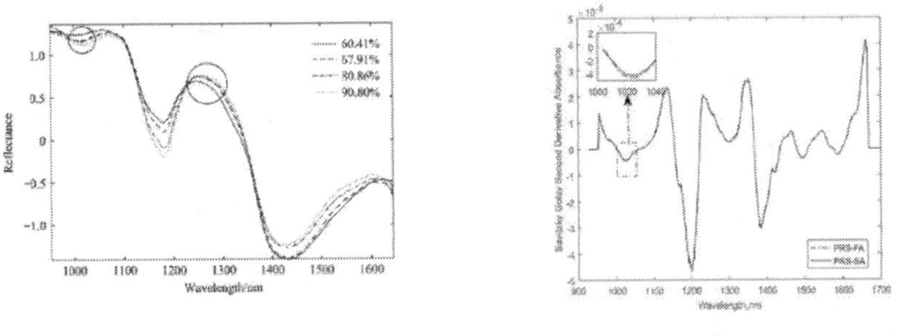

(a) (b)

Figure 2. (a) Average reflectance spectra of the cup lump rubber for different levels of DRC and (b) the spectral data preprocessed with the second derivative of Para rubber sheet for transflectance mode.

3.2 Classification of adulterated Para rubber sheet

The second derivatives of the spectral data, which resolved peaks in the spectra, are displayed in Figure 2(b) for transflectance mode. Strong peak at 1200 nm was observed, which was assigned to C-H stretching

IOP Conf. Series: Materials Science and Engineering 1234 (2022) 012037 doi:10.1088/1757-899X/1234/1/012037

second overtones or part of isoprene [3]. The model based on the transflectance mode yielded good classification performance in terms of R^2 of 0.88, standard error of prediction (0.88) and accuracy of 98.33%. For the purpose of visualized classification, the HSI image was mapped into color with the class value on each pixel predicted by the best model and showed in Figure 3.

3.3 Development of NIR portable system for quality evaluation of Para rubber timber

The timber samples contained the moisture content in a range 2.59% to 88.59%. The PLSR model gave good performance in cross validation providing Rcv of 0.98 and root mean square error of cross validation of 5.48%. The portable NIR spectrometer (Figure 4) was developed based on the predictive model and could be used to non-destructively and rapidly measure the moisture content in Para rubber timber with acceptable accuracy.

Table 1. Predictive performance on DRC based on different preprocessing.

Preprocessing method	R^2	RMSEP	RPD
SNV	0.99	0.90	12.10
SNV+first derivative	0.99	0.74	14.69
SNV+second derivative	0.99	0.64	16.83

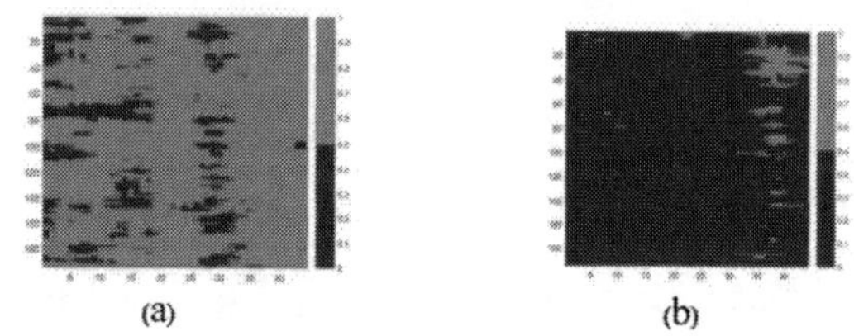
(a) (b)

Figure 3. Color-mapped images for (a) formic acid, and (b) sulfuric acid.

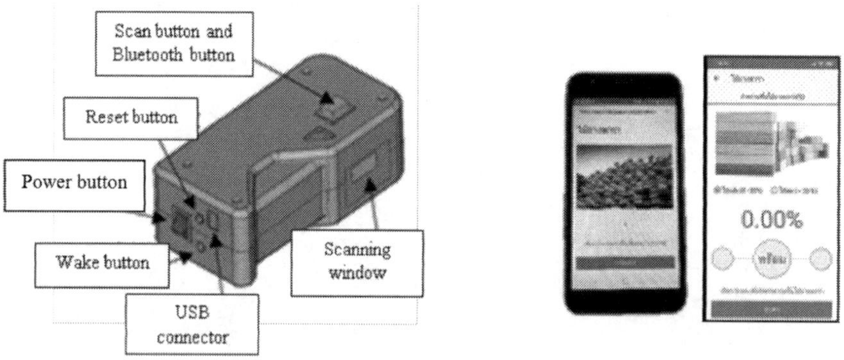

Figure 4. (a) Developed hand-held system for measuring the moisture content in timber and (b) android application.

4. Conclusions

The NIR has benefit of being fast and non-destructive technique. NIR limitation is light penetration depth into the material, which allows detection depth of about 5-10 mm from the sample surface. Our research group has applied NIR successfully in determination of DRC in rubber cup lump, classification of adulterated rubber sheet and development of a portable system for measuring moisture content in rubber timber.

However, the results could be implemented in development a commercial instrument for quality control in near future.

Acknowledgement
The authors gratefully acknowledge the Thailand Research Fund (Grant number: RDG62T0001) and the Faculty of Engineering at Kamphaengsaen, Kasetsart University, Thailand for financially support of the research.

References

[1] Lu R F and Chen YR 1998 Hyperspectral imaging for safety inspection of food and agricultural products. In Proceedings of SPIE Conference on Pathogen Detection and Remediation for Safe Eating, Boston, US,
[2] Williams P 2014 Tutorial: The RPD statistic: A tutorial note. NIR News, 25(1): 22.
[3] Osborne BG, Fearn, T and Hidle, PH 1993 Practical NIR Spectroscopy with Applications in Food and Beverage Analysis.

AUTHOR INDEX